BEEKEEPING STUDY NOTES

for the

BBKA BASIC
SBKA BASIC BEEMASTER
FIBKA PRELIMINARY

EXAMINATIONS

Prepared by:

J.D.YATES B.Sc.(Hons), C.Eng., FIEE.
and
B.D.YATES SRN, SCM.

Northern Bee Books

BEEKEEPING STUDY NOTES

© J D Yates and B D Yates

ISBN 978-1-912271-03-0

First published 1999

Reprinted 2012

Published by Northern Bee Books, 2017
Scout Bottom Farm
Mytholmroyd
Hebden Bridge HX7 5JS (UK)

Printed by Lightning Source, UK

PREFACE

We have been asked many times why we have not prepared notes for the Candidate contemplating the BBKA Basic Examination when we have catered for all the other BBKA examinations in our other series of study notes. We had no answer and believe now that this has been a great omission on our part.

We rate very highly the BBKA examination system and believe that all beekeepers should study and attempt this very simple Basic Examination to provide them with the confidence to manage their bees in a safe and competent manner. This applies to those beekeepers both within and without the BBKA. Within the BBKA, we believe that it should be mandatory for all members to attempt the Basic Examination during the first three years of becoming a member; something for the Executive to mull over. It would be an act of foolhardiness to set sail for France without some formal training in seamanship and navigation; similarly, we regard it as an act of foolhardiness to keep bees without some formal training in the basics of beekeeping husbandry. The BBKA Basic Syllabus and Examination provides this training.

The contents section of our notes is the syllabus for the BBKA Basic Examination and each part of this syllabus is addressed in the following notes which, hopefully, cover everything that the Examiner is likely to ask the Candidate. The Scottish and Irish syllabi are very similar and any additional syllabus items for their national examinations have been addressed in separate appendices, Appendix 6 for the Scottish Basic Beemaster Certificate and Appendix 7 for the Preliminary Certificate of the Federation of Irish Beekeepers' Associations.

Many candidates enter the examination not knowing what to expect; have no further fears. The examination is a very simple one in four parts; assembly of a frame and fitting it with foundation, a practical manipulation at the hive and an oral in two parts, one part on diseases and pests (section 5.0) and the other part on general beekeeping knowledge (sections 1.0 to 4.0). During the colony manipulation of the brood chamber questions will be asked about eggs, brood, food, etc. readily observed on the combs removed from the hive. Your confidence and competence handling the bees will be noted by the Examiner. This part of the examination takes about 15 to 20 minutes. The whole examination is complete in an hour to an hour and a half. The Examiner will take note of the weather conditions and the temperament of the bees and make allowances as necessary. Take heart that all examiners will be doing their level best to pass the Candidates.

Passing the Basic is the key which opens the door to understanding honeybees and their associated husbandry through the modular system of examinations (shown in Appendix 5) administered by the BBKA's Examination Board. We hope you enjoy the experience; Examiners try to make it a pleasant one.

Finally, we express our thanks to John Annett of the Harrogate and Ripon Beekeepers' Association for his diligent proof reading of the manuscript and finding all those little 'funnies' lying in the text before handing it over to the typesetters.

JD & BD Yates,
Newton Ferrers, Devon. 1999

CONTENTS

1.0 MANIPULATION OF A HONEYBEE COLONY

The Candidate will be aware of:

1.1 the care needed when handling a colony of honeybees,
1.2 the reactions of honeybees to smoke,
1.3 the personal equipment needed to open a colony of honeybees,
1.4 the reasons for opening a colony,
1.5 the need for stores.

The Candidate will be able to:

1.6 open a colony of honeybees and keep the colony under control,
1.7 demonstrate the use of the smoker,
1.8 demonstrate the use of the hive tool,
1.9 remove combs from the hive and identify worker, drone and queen cells or cups if present, and comment on the state of the combs,
1.10 identify the female castes and the drone,
1.11 identify brood at all stages,
1.12 demonstrate the difference between drone, worker and honey cappings,
1.13 identify stored nectar, honey and pollen,
1.14 take a sample of worker bees and put them in a match box or similar container,
1.15 state the number of worker bees required for an adult disease diagnosis sample,
1.16 demonstrate how to shake bees from a comb and how to look for signs of brood disease.

2.0 EQUIPMENT

The Candidate will be:

2.1 able to name the principal parts of a modern beehive,
2.2 aware of the concept of the bee space and its significance in the modern beehive,
2.3 able to assemble a frame and fit it with wax foundation,
2.4 aware of the reasons for the use of wax foundation,
2.5 aware of the spacing of the combs in the brood chamber and super for both foundation and dra comb and methods used to achieve this spacing.

3.0 NATURAL HISTORY OF THE HONEYBEE

The Candidate will be:

3.1 able to give an elementary account of the production of queens, workers and drones in the honeybee colony,

3.2 able to state the periods spent by the female castes and the drone in the four stages of their life (egg, larva, pupa and adult),

3.3 able to give an elementary description of the function of the queen, worker and drone in the life of the colony,

3.4 able to give a simple description of wax production and comb building by the honeybee,

3.5 aware of the importance of pollination to flowering plants and consequently to farmers and growers,

3.6 able to name the main local flora from which honeybees gather pollen and nectar,

3.7 able to give a simple definition of nectar and a simple description of how it is collected, brought back to the hive and is converted into honey,

3.8 able to give a simple description of the collection and use of pollen, water and propolis in the honeybee colony,

3.9 able to give an elementary description of swarming in a honeybee colony,

3.10 able to give an elementary description of the way in which the honeybee colony passes the winter.

4.0 BEEKEEPING

The Candidate will be:

4.1 able to give an elementary description of how to set up an apiary,

4.2 able to describe what precautions should be taken to avoid the honeybees being a nuisance to neighbours and livestock,

4.3 describe the possible effects of honeybee stings and able to recommend suitable first aid treatment,

4.4 able to give an elementary description of the year's work in the apiary and of the management of a colony throughout a season,

4.5 able to describe the preparation of sugar syrup and how and when to feed bees,

4.6 aware of the need to add supers and the timing of the operation,

4.7 able to give an elementary account of one method of swarm control,

4.8 able to describe how to take a honeybee swarm and how to hive it,

4.9 able to describe the signs of a queenless colony,

4.10 able to describe the signs of laying workers and of a drone laying queen,

4.11 aware of the dangers of robbing and how robbing can be avoided,

4.12 able to describe one method of uniting colonies,

4.13 aware of the reasons for uniting bees and the precautions to be taken,

4.14 able to describe a method used to clear honeybees from supers,

4.15 able to describe the process of extracting honey from combs and a method of filtration and bottling of honey suitable for a small scale beekeeper,

4.16 aware of the need for good hygiene in the handling of honey for human consumption,

4.17 aware of the legal requirements for the labelling and sale of honey,

4.18 able to give an elementary account of the harvesting of beeswax.

5.0 DISEASE, POISONING AND PESTS

The Candidate will be:

5.1 able to describe the appearance of healthy brood and how it differs from diseased brood or chilled brood,

5.2 able to describe the signs of the bacterial diseases American Foul Brood (A.F.B.) and European Foul Brood (E.F.B.) and describe their effect upon the colony,

5.3 able to describe methods for detecting and monitoring the presence of *Varroosis* (a mite) and describe its effect on the colony.

5.4 aware of *Acarapis* (a mite) and *Nosema* (a protozoa) and their effect upon the colony,

5.5 able to describe ways of controlling Varroosis using one registered product and one recognised biotechnical method.

5.6 aware that *Braula coeca* is neither a mite nor a parasite, but is an insect that steals food,

5.7 able to distinguish between *Varroa jacobsoni* and *Braula coeca*,

5.8 aware of the current legislation regarding notifiable diseases of honeybees,

5.9 aware of the national and local facilities which exist to verify disease and advise on treatment,

5.10 know where to obtain assistance if any poisoning by toxic chemicals is suspected,

5.11 able to describe how comb can be stored to prevent wax moth damage,

5.12 able to describe how mice and other pests can be excluded from the hives in winter.

APPENDICES

1. Migration and evolution of the honeybee.
2. Food sharing in the honeybee colony.
3. Colony development during the active season.
4. Honey usage diagram.
5. The BBKA examination system.
6. Additional items to cover the Scottish Basic Beemaster Certificate.
7. Additional items to cover the Federation of Irish Beekeepers' Associations Preliminary Certificate.
8. The SBKA examination structure.
9. Additional reading extracted from the BBKA book list.

** ** ** **

FOREWORD

At last John and Dawn Yates have written the study notes for the BBKA Basic Examination. This text comprehensively covers all the requirements of this examination together with appendices to cover the additional requirements in the equivalent examinations arranged by the Scottish Beekeepers' Association and the Federation of Irish Beekeepers' Associations.

This book provides excellent practical advice and help for those contemplating taking their first beekeeping examination of proficiency in apiculture. Potential candidates should be aware that the standard required is that of a beekeeper who has kept and managed bees in a safe and efficient manner for one season. Candidates should be able to demonstrate good skills with the smoker and use of a hive tool and are encouraged not to wear gloves, but those who elect to wear gloves for the practical part of the examination will not be penalised.

It is not possible for an examiner to cover all the detail covered in this book in an examination only lasting an hour or so and the examiner will select the appropriate points on which to ask questions. It is important that candidates are aware of principles covered by the syllabus and that is where this book will be invaluable in explaining this theory to the student.

Students should not neglect other books on the recommended list and a sensible approach would be to select one or two others at this stage. This will then start a library which can be used to proceed to the modules and other examinations. Reading more than one book will give a broader view to the subject.

All beekeepers should take the Basic Examination after one or two seasons beekeeping. Success at this level will give them encouragement to become better beekeepers.

A text full of useful information presented in a logical order the book will also help those designing courses for the Basic examination.

John Hendrie
26, Coldharbour Lane,
Hildenborough,
Kent

1.0 MANIPULATION OF A HONEYBEE COLONY

This section describes the way we handle our own colonies of honeybees; as you become more experienced you will probably develop your own techniques.

1.1 The care needed when handling a colony of honeybees.

Honeybees are primitive insects and they are much the same as they were 10 to 20 million years ago, wild and untamed. You will probably have seen in your reading reference to the domestic bee as opposed to the bumble bee; the honeybee has not been domesticated and is unlikely to be. Man has only learnt, mainly through observation, trial and error, how to manipulate honeybees for his own purposes.

1.1.1 Great care must be exercised when opening a colony in order to:

- ensure your own safety.
- ensure the safety of others (neighbours and passers-by).
- ensure the safety of pets, any other domestic animals and livestock.
- disturb the activities of the bees within the hive during the inspection as little as possible.

1.1.2 The temperament of honeybees can range from being very docile to very aggressive. Understanding the reasons for this wide range of reaction to intrusion into their nest is helpful in determining our approach:

- The honeybee's temperament is part of its genetic make up and in no way can it be improved except, of course, by selective breeding.
- Bees are more aggressive during bad weather eg. rain and electric storms (thunder and lightning). Bees are known to be irritable if stocks are sited under power lines so they probably have the ability to measure electrostatic charge. The orientation to the earth's magnetic fields is known to have an affect on comb building.
- Queenless colonies are also more aggressive than queenright colonies.
- Good tempered bees can become aggressive very quickly (during a manipulation).
- Bad handling can, and certainly does, make bees bad tempered.

From the above list it would seem sensible to enlarge on the genetic causes of temperament and the bad handling aspects, points of which the beginner and those attempting the examination should be aware.

1.1.3 The honeybee's temperament is built into its genes and can range from the very docile to the extremely aggressive. The very docile honeybee can be handled by almost anyone without smoke or personal protection whilst the very aggressive colony can defy the best efforts of even the expert bee handler. It is worthwhile listing the characteristics which can be found in different types of colony in the United Kingdom. These characteristics are usually referred to, incorrectly, as the temper of the honeybee. Against this background we are referring to those characteristics which manifest themselves when a manipulation of the colony is undertaken or a predator (beekeeper, neighbour or animal) is within the vicinity of the nest site; a direct intrusion into the nest of the honeybee or its environs. This defensive behaviour of the bees is usually measured by the stinging propensity of the colony concerned.

a) Guarding around the nest site. All honeybees will guard their nest site, some only at the entrance while at the other extreme are those bees which attack an intruder at large distances from the nest site. The Africanised honeybee of South America will guard at up to 500 meters from its undisturbed nest site and can be extremely dangerous. In the United Kingdom the worst situation is 50 to 100 metres for an undisturbed colony. Such bees are quite unacceptable and should be destroyed. Honeybees which follow to such ranges after a colony manipulation are equally unsatisfactory in this country and should be requeened or destroyed. After a manipulation of the colony it should be possible to remove one's veil at a distance of 2 metres to the sides and rear of the colony without bees buzzing around one's head; these bees are generally referred to as 'followers'. Honeybees which follow to greater distances than this are totally unsuitable for keeping in a suburban garden. In an apiary, followers stay with the beekeeper when moving to the next hive to be inspected. When this hive is opened it is these following bees that 'upset' the bees in the colony, very often making them more difficult to handle. Our experience indicates that the larger the following distance the greater the propensity to sting.

b) Runners. These bees vacate the combs they are covering as the colony is inspected leaving the brood exposed and the combs bare of bees. When the bees react like this it is very difficult to find the queen, the bees may run to the bottom corner of a comb as it is removed from the hive and then drop off in one cluster, down your 'wellies' if you are not holding the frame over the open hive. At the other extreme the bees can be very quiet on the comb as it is removed and cover the brood completely all the time the comb is out of the hive. The hyperactive bees are more inclined to sting than bees which remain quietly on the comb.

c) Nervousness. This characteristic of the bees is demonstrated when the frame is removed for inspection and the bees appear agitated, continually moving over the surface of the comb; they do however still cover the brood. Generally they are prone to sting and require careful handling with slow deliberate movements. Manipulating using cover cloths can be helpful.

d) Reaction to smoke. Most bees react favourably to smoke; they are repelled by it and gorge themselves with honey. In a few cases some colonies react unfavourably and become more aggressive the more they are smoked. Bees with such a temperament are very prone to sting and very difficult to handle.

The above characteristics can be found to a greater or lesser degree in most mongrel bees found in the United Kingdom. Our observation of the temperament of honeybees in this country leads us to believe that nothing has changed during the last 50 years and the temper of the mongrel bee is just as bad as it was all those years ago. BIBBA (Bee Improvement & Bee Breeders' Association) will state otherwise and claim that the introduction of other strains from overseas is responsible for bad temper.

Bad temper is easily dispensed with by culling queens producing bad tempered progeny. The real trouble, in our opinion, are the many beekeepers who do not follow this simple practice.

1.1.4 Bad handling techniques can be corrected by the beekeeper. The points to avoid are as follows:

- Any vibrations, bangs, knocks, etc.
- Any fast, quick or speedy movements.

Initially, when learning to handle bees, go slowly, making all actions deliberate and purposeful.

1.1.5 It is necessary to be able to detect any signs of restlessness or change in the colony's behaviour during a manipulation. Points to watch for are as follows:

- Activity at the entrance is the first indicator. Study this before touching or smoking the colony so that you have a mental picture of the normal activity of the bees at the hive entrance.
- Watch the entrance regularly from time to time while manipulating the colony; this must become second nature to all beekeepers.
- Any bees starting to collect at the entrance or starting to collect up the front of the hive is a warning sign of trouble. Other indications of increasing irritability of the bees within the hive are:

  a) Rows of eyes between the frames spells trouble; rows of 'bottoms' means a contented colony.
  b) Bees darting off the frames (vertical take off), while the frames are being removed or replaced.
  c) A change in the sounds coming from the bees within hive.
  d) A change in the behaviour of the bees i.e. not remaining quietly on the combs.
  e) A smell of banana oil (an alarm pheromone called isopentyl acetate). However, please note that we have only come across one beekeeper with this remarkable attribute of being able to detect this smell while manipulating a colony. We have only noticed it in the car while moving bees.

1.1.6 One final word on the need for care when handling a colony of honeybees. If the colony, in your opinion, is becoming uncontrollable then close it down immediately. You will not lose face by doing this; it is good beekeeping practice. We have seen, so many times, demonstrations where the colony is out of control, observers are being stung unnecessarily and the demonstrator still pressing on; bravado such as this is, in our opinion, just sheer stupidity.

1.2. Reaction of honeybees to smoke.

1.2.1 Providing the smoke is cool and has a pleasant wood smell (a somewhat subjective statement) honeybees usually move away from the smoke and immediately and instinctively gorge themselves with honey. With full honey sacs they are in a docile state. The origin for this is obscure, but the popular story is absconding of the colony in the wild state at the onset of a forest fire. Generally honeybees move away from smoke so that smoking the top bars and frame lugs during a manipulation will keep the bees below the top bars of the frames. Similarly, smoke will drive the guard bees into the hive from the hive entrance.

1.2.2 Smoke is thought to mask the odours within the hive preventing chemical messages being transmitted amongst the bees and masking the odours of intruders be they bees or beekeepers. A recent study (1995 by Vischer, Vetter and Robinson) found that the sensitivity of the antennal receptors was markedly reduced after being subjected to smoke. It is still not clear how smoke affects the honeybee making it more docile and easy to handle. J.B.Free showed that smoked bees remained gorged for about two hours but began to sting again after 15 minutes unless subjected to further smoke.

1.2.3 There are many acceptable fuels that can be used in the smoker and every beekeeper seems to have his favourite:

• Wood shavings from a plane are ideal for starting the smoker; they can be lit directly, like paper, with a match.
• Softwood planer chippings from a sawmill or timber yard are a good fuel.
• Dry rotted wood is good and can be broken up easily by hand; it possibly burns a bit on the hot side and leaves a sticky residue in the smoker.
• Dry grass cuttings are an old favourite. The coarser the grass the better allowing a natural air flow through the smoker.
• Sacking or hessian burns steadily with few sparks.
• Corrugated paper rolled into a cartridge used to be popular but much of this packing material is now impregnated with chemicals to prevent it burning. Some beekeepers interlaced the corrugated cardboard with hessian.
• Old oily rags give off copious black smoke.

1.2.4 Some thoughts on the smoker (the beekeeper's most important tool):

• Purchase the largest you can afford.
• Preferably in copper or stainless steel.
• Have spare fuel and lighter to hand during a colony manipulation.
• Use a fuel which burns slowly and steadily producing a cool spark free smoke.
• Before starting a manipulation ensure that the smoker is fully fuelled and stays alight.
• In our opinion, there is only one place for the smoker during a manipulation and that is between your KNEES; make this practice second nature to you when manipulating, the smoker will then always be in the right place and ready for use quickly if needed.
• Keep it in your possession throughout the manipulation.
• Extinguish the smoker when you have finished ie. plug the nozzle with green grass or cork.
• Be careful not to start a fire by carelessness.

1.2.5 Note our comments in 1.1.3 that the occasional colony reacts adversely to smoke, aggravating the bees instead of placating them.

1.3 The personal equipment needed to open a colony of honeybees.

The following equipment is required by every beekeeper:

- A good quality veil, hive tool and smoker are essential.
- The following are optional:

    a) Gloves, bee suit, 'wellies', etc.
    b) A couple of cover clothes can be very useful.

Because bees are often handled very badly and because insufficient effort is made by beekeepers to cull for bad temper and rear docile bees, beekeepers protect themselves thoroughly and often are responsible for innocent bystanders getting stung. PLEASE PONDER THIS POINT. We remember, on one occasion, parking our car at the side of the road near a hedge and, on opening the door and getting out, one of us was stung twice on the face in as many seconds. Further investigation revealed a beekeeper in his apiary on the other side of the hedge and out of sight from the road. We shudder to think what could have happened to an allergic child. The beekeeper, of course, was dressed up to the 'nines'.

We are firmly of the opinion that if your bees cannot be handled without gloves and 'wellies' then there is something wrong with the bees or the handling techniques of the beekeeper; both can be readily corrected.

Honeybees have very sensitive sense cells in their antennae. Bees are attracted to flower-like aromas which indicate to the bees a source of nectar. Volatile synthetically produced aromas contained in some hair sprays, hair shampoos, perfumes etc. also attract bees because of their similarity to floral aromas. When bees are alarmed they give off a pheromone (a chemical message) alerting all the bees in the hive. This chemical message has a banana-like odour similar to an essence used in some commercially produced perfumes contained in products produced to enhance or disguise the odours of mankind. A beekeeper should be odourless. This is achieved by good personal hygiene and avoiding confusing his/her bees by not using any scented soap, perfume etc. prior to inspecting stocks of bees.

1.4 The reasons for opening a colony.

1.4.1 In a branch teaching apiary, stocks of bees are frequently opened unnecessarily and in unfavourable conditions in order to demonstrate certain aspects of the honeybee colony for newcomers to the craft of beekeeping. Opening a stock of bees to inspect the contents of the removable frames completely disorganises the activities of the bees within the colony. As a consequence of exposing the contents of the hive for inspection, bees will waste their energies on fanning to remove the smoke, heat will be lost from the brood nest, cleaning up the remains of squashed bees will involve additional work by the house bees, returning foragers will be disorientated by the presence of the beekeeper and pollen loads are often abandoned instead of being packed neatly into storage cells.

1.4.2 To avoid stress to the bees a stock should only be opened for a particular reason. The manipulation or inspection should be carried out ideally on a warm sunny day with little or no wind and with a minimum ambient temperature 60°F (15.5°C). The best time to open a strong stock is when most of the foragers are out collecting nectar and pollen from the blossoms, this is said to be about mid-day. The inspection should be carried out efficiently with the minimum of disturbance, the hive being opened for the shortest possible time. Stocks of bees are inspected during the active season in order to:

5

- Check on the presence of a fertile queen.
- Check on the health of the colony and its expansion.
- Undertake any necessary swarm prevention measures.
- Check for any sign of disease either in the brood or adult bees.
- Check that the colony has sufficient stores.
- Check that the colony has sufficient space for bees, for the queen to lay and for the bees to store honey.

1.4.3 Five points require consideration each time a colony is opened during the active season:

1. Relating to the queen. Is the queen present? Is she seen during the inspection, is she marked so that she can be easily identified? Have the wings been clipped? Is she laying viable worker brood? Is there the correct proportion of eggs to open brood to sealed brood present? Are there any queen cells present? Is there plenty of space for the queen to continue to lay? These observation are important for the control of swarming which is dealt with in sections 4.7 and 4.8.

2. Relating to the brood. Has the brood increased or decreased since the last inspection? If the queen is laying 1,000 eggs every 24 hours, then one would expect to notice an increase in the number of frames containing brood during the spring build up of the colony. A British National frame has 2,500 cells on each side of the frame; only the centre part of the frame would be used for brood (about 75%), the cells in the top arches of the frames being used to store pollen and honey (about 25%). Thus it would take the queen about 4 days to fill one frame with brood. Are the empty cells polished and ready for the queen to lay her eggs? Is there only one egg to a cell? Is there any drone brood? Are there sufficient bees to cover the brood?

3. Relating to disease. Firstly, do the adult bees look healthy, ie. without deformity or parasites? Is there any soiling of the combs or hive furniture? Do the bees move normally on the frames? Are there any dead bees outside the entrance or on the floor board? Secondly does the brood appear healthy, ie. open brood pearly white, sealed brood without any perforated or distorted cappings? To look for any sign of European Foul Brood (E.F.B.) or American Foul Brood (A.F.B.) see section 5.2.

4. Relating to stores. During the summer months a strong colony may use up to about a pound of honey each day. This is consumed by the bees to produce energy for heat, mobility and for the bee to maintain all its bodily functions necessary for the survival of the colony eg. brood food, wax, glandular secretions. So for every day between inspections there should be at least a pound of honey stored on the frames. Stored pollen should also be present. These stores will then allow the colony to continue to thrive during adverse weather conditions and a dearth of forage until the next inspection.

5. Relating to space. As the colony grows throughout the season, reaching its peak of population in July, room must be available within the hive for the increasing numbers of bees and for the honey stores. As soon as the bees appear to be covering 8 or 9 frames of bees within the brood chamber then a queen excluder and the first super should be added. This additional space allows the brood box to be used almost entirely for brood. Should there be any frames which are pollen

clogged or full of granulated stores, they are best removed and replaced with frames of comb or foundation. Depending on the season and available forage, bees can collect so much pollen that it is stored on combs at the side of the brood nest; such combs become a barrier preventing the natural expansion of the brood nest. Similarly, frames of sealed honey stores which are no longer in the liquid state, ie. granulated stores associated with nectar collected from the blossoms of the brassica family, should be removed.

Subsequently further supers must be added when 8 frames in the top super are covered with bees. These observations should be recorded on a hive record card each time the stock of bees is inspected as well as in a log book. The hive record card is best kept in the roof of the hive. The information in the log book kept at the beekeeper's home is invaluable to determine equipment required for the next colony inspection.

1.5 The need for stores.

1.5.1 The forager bees collect water, propolis, pollen and nectar. Bees forage during daylight hours when the temperature is above 50°F (10°C), when the wind does not exceed 40 mph and when the flower blossoms are open and producing nectar and pollen. Most of the foraging flights undertaken by bees occur during the months April until September, depending on the climate of the district, availability of forage and the shelter provided by a well-chosen site for the apiary. From October through to the end of March the colony of bees will be depending on the stores of honey and pollen preserved by the bees within the hive during the previous active season, supplemented by the small amounts of freshly collected water, pollen and nectar when the weather is favourable. There are always short spells of sunny weather during the winter period when bees are able to leave the hive for cleansing flights and to forage for short distances from the hive. Propolis needs to be soft and pliable for bees to collect; some of the earliest propolis available to bees in the spring is collected from the sticky buds of the Horse Chestnut tree, *Aesculus hippocastanum*. Exotic well-sheltered plants eg. Honeysuckle, *Lonicera standishii* or Daphne, *Daphne mezereum* bloom in February together with the crocus, *Crocus spp.* and snowdrop, *Galanthus nivalis* and may yield some nectar and pollen but most of the flowers of plant species native to this country require a warm ambient temperature for the plants' nectaries to secrete eg. 18°C (64 °F) for Cherry Laurel, *Prunus laurocerasus* (Floral Biology by M.S.Percival).

1.5.2 Water is only collected as required by the colony. It is not stored.

1.5.3 Propolis is collected as required by the colony. Different species of bee are more prone to collecting 'excess' propolis eg. Caucasian bees from the Eastern Mediterranean. Propolis is used by the bees to seal up small cavities, to strengthen the wax comb and to varnish the inside of brood cells prior to the queen laying in them. It is not deliberately stored for times of dearth.

1.5.4 Pollen provides the proteinaceous part of a bee's diet. It is fed to the open brood, ie. larvae from the 3rd day, and eaten by the young nurse bees thereby stimulating the development of the exocrine glands of the bee, eg. hypopharyngeal glands which produce brood food. The presence of open brood within the brood chamber produces a pheromone which stimulates bees to forage for pollen. 70-150mg of pollen is required to produce 1 bee. It is thought that during the active season a large colony of bees will collect and consume as much as 50-100lbs (23-46kg) of pollen.

1.5.4.1 Pollen is usually stored in an arch around the brood nest beneath the stores of sealed honey. There are exceptions to this, for example when the bees are working rape, *Brassica napus,* there is so much pollen available that pollen is deposited in the brood combs even occupying all the cells of one or both sides of a frame. Ivy, *Hedera helix,* is another prolific provider of pollen. The 'house' bees pack the pollen into the cells, mix honey with the pollen (pickled pollen) then finally, when the cell is full, the house bees seal the cell with a layer of wax. Pollen which is not properly stored becomes infested with fungal spores and becomes useless to bees as a food source.

1.5.4.2. During the period when little fresh pollen is available, October through to March, stored pollen is used. Colonies such as those sited in the grain producing fields of the South Eastern areas of UK where the hedges and trees have been destroyed to allow easier harvesting of crops, are likely to be short of pollen. Beekeepers can remedy this shortage by placing pollen 'patties' directly over the top bars of stocks of bees in early spring.

1.5.4.3 Candidates should be able to differentiate between:

  • the sealed cells of brood, stored honey and the sealed cells of stored pollen. Pollen and honey cells will be sealed with wax whilst the brood cells are sealed with a mixture of wax and pollen. The honey cells will be on the periphery of the brood frames with a light coloured capping. The pollen cells will be found between the brood and the honey stores and usually have a darker coloured wax capping. The brood cells have a convex capping of old wax mixed with pollen in order that the capping is porous to allow the larva to breathe. Experiment by partially opening a sealed cell with the blade of a small pocket knife until you are sure of the difference.
  • cells of pollen invaded by the fungus, *Bettsia alvei,* and cells of brood infected by the spores of Chalk Brood, *Ascosphaera apis.* When the bees clean out the contents of these cells, leaving the debris at the entrance of the hive, a close inspection is needed to differentiate between the layered consistency and colour of the infested pollen and the granular texture and colour of the larvae infected with chalk brood.

1.5.5 Honey is processed from nectar. Nectar is a sugary solution secreted by the nectaries of flowers. When the honey is 'ripe' it is sealed in cells in the wax comb by a capping of wax. This honey is now approximately 20% water and 80% sugar; it will not ferment or deteriorate. When the colony is short of incoming nectar, the cappings are opened and, the honey being hygroscopic, will absorb water and will be diluted and digested by the bees as a 50% water and sugar solution. Stores of honey provide the bees with:

  • Sugar which is converted into energy for activating all the body processes necessary to sustain life.
  • Water which keeps the sugars in solution.
  • Organic acids, basic elements, amino acids and about 0.2% of protein all of which are necessary for the well-being of the bee.

A strong stock of bees during the active season, March - September, may collect, use and store for the colony's own consumption as much as 350 pounds (159 kilograms) of honey. These stores are used by the bees for:

- Fuel for the foraging bees.
- Providing the fuel for heating the brood nest or the cluster of winter bees.
- Food for the older larvae and adult bees.
- Indirectly providing the brood food for the young larvae and royal jelly for the queen. These secretions are produced by the mandibular and hypopharyngeal glands of the worker bee after consumption of stores.
- By the continual interchange of food between the hive bees, trophallaxis, a communication system is in constant use, signalling the presence of the queen and the state of the hive's resources.

1.5.5.1 Stocks of bees require feeding sugar syrup when the stores of honey are inadequate in order to prevent the bees dying from starvation. Shortages of stores occur when:

- 'Surplus' honey stores are usually removed after the main 'flow' at the end of July. In return the beekeeper must replenish the colony with sufficient stores (sugar syrup) to survive the winter ie. 35-40 pounds (16 - 18kg) of sealed liquid stores.
- Spring honey is sometimes removed from the bees in May. Honey derived from the nectar of oil seed rape, sown the previous autumn, needs to be removed soon after the honey is sealed in the combs. If left until later the honey granulates in the comb and cannot be spun out of the combs by the beekeeper or used as food by the bees. The removal of spring honey may necessitate feeding the bees sugar syrup to replace these stores unless the weather is favourable and there is ample nectar available for the foragers to collect. During May the colony is still expanding and may be consuming about 1lb (0.45kg) of honey a day to provide food for the rapidly increasing population of bees.
- When nuclei are made up during the queen rearing season there must be adequate stores for the small colony to survive until it produces its own foraging force of bees. Feeding is recommended after three days, to prevent robbing, if the nucleus remains in the same apiary as the donor hive. This reduces the risk of robbing as the foraging bees will all have returned to the original site. Therefore always make up a nucleus hive with a frame of stores.
- A swarm of bees usually brings with it sufficient stores to establish a new brood nest. A period of poor weather when the bees are unable to forage for nectar may bring the newly-hived swarm to near starvation. Creating the new comb uses a great deal of honey, producing 1 pound wax uses 6-8 pounds (2.7-3.6kg) of honey.
- Any prolonged period of cold, wet and windy weather when bees are unable to forage may mean starvation to a hive of bees unless there are adequate stores of honey. Stocks of bees are particularly vulnerable during the months of March and April when the brood nest is expanding rapidly and the brood outnumbers the adult bees. Similarly, during the month of June, when the hive is approaching its maximum population of bees, stocks of bees sited away from any suitable forage combined with a period of unfavourable weather will rapidly use all reserves of stores and require feeding.

1.6 Opening a colony of honeybees and keep the colony under control.

A colony can only remain under the control of the beekeeper if:

• The bees are in the hive.
• The colony has been subdued with smoke.

1.6.1 When opening a colony for, say, a routine inspection the following sequence should be followed:

a) Approach the colony quietly and gently place any spare equipment on the ground near the hive.
b) Observe the normal activity at the entrance.
c) Smoke at the entrance (6 good puffs of the smoker). Don't be timid with the smoker at this stage.
d) Wait for at least two minutes (you should know why).
e) Gently remove the roof and place it on the ground, upside down, behind the hive with a corner pointing to the back of the hive. This may sound a bit pernickety but it is now all ready in the right position to put the supers on.
f) Now study the record card, check that you have all the equipment required and that you are sure what has to be done.
g) Two small puffs of smoke at the entrance; we're coming in now!
h) Ease the supers up on one side with the hive tool, tilt and smoke well between the supers and the queen excluder.
i) Lift off supers and place gently on the upturned roof. No bees will be squashed as there are only 8 small points of contact.
j) Smoke the bees down so that the queen excluder is clear of bees and then gently lever and twist it off. Hold it over the hive and shake any remaining bees off and place behind the hive.
k) NOW PUT THE SMOKER BETWEEN YOUR KNEES ready for use.
l) Remove the dummy board, shake off any bees into the top of the hive with a sharp tap and lean it up at the rear of the brood chamber with the queen excluder.
m) A small puff of smoke to clear the bees at the lugs of the first frame (where you want to hold them) and gently remove it, inspect and replace it hard up against the side of the brood chamber.
n) Don't forget to glance at the entrance.
o) If the bees are tending to well up and fly off, smoke them down.
p) Repeat the operation with every frame always keeping the bees down in the hive with the smoker. No jerking or bumping or fast movements are acceptable.
q) Replace the dummy at the opposite end of the brood chamber and lever the frames and dummy with the hive tool so that there are no gaps between the spacing devices. This will prevent a build up of propolis which would change the frame spacing.
r) Smoke the tops of the frames clear of bees and replace the queen excluder.
s) Replace the supers carefully without squashing any bees.

Note: if there are no supers on the colony, check that the queen is not on the underside of the crown board when it is removed.

1.6.2 If the bees are difficult to keep down in the hive with smoke then use cover cloths, but avoid them if possible. We have found that sunlight/daylight also tends to subdue some strains of bee. If cover cloths are needed most of the time then, in our opinion, the strain of bee could usefully be changed.

1.6.3 THERE IS AN ACQUIRED ART IN THE USE OF SMOKE; WATCH CAREFULLY A COMPETENT BEEKEEPER AT WORK, THE ART IS EASY TO LEARN.

1.6.4 If you feel the colony is getting out of control, then close it down before you get into a 'tail spin'. It is the most sensible thing to do and is good beekeeping practice.

1.6.5 It is preferable not to wear gloves to manipulate your colonies, have them by you and on standby for an emergency. If you are frightened of a sting or you are allergic to stings, then beekeeping is not your forté.

1.7 Demonstrate the use of the smoker.

1.7.1 During the examination, the part where the Candidate has to manipulate the colony, the Examiner will be noting how you use the smoker. First you will be asked to light it. A seemingly simple task but Candidates have been failed for not being able to accomplish the lighting procedure. It is no good filling the smoker with fuel and throwing a lighted match on top; it will go out no matter how hard the bellows are worked. The burning fuel must be at the bottom of the fire box. You should be able to light the smoker in the apiary with a breeze blowing. We believe that using a small gas blow torch to be a doubtful starter; no pun intended!

1.7.2 It is important that the smoker is burning well, full of fuel and producing plenty of cool smoke before approaching the hive. To achieve this state of affairs will take a few minutes to produce some hot ashes at the bottom of the fire box necessary to keep the smoker burning by itself without pumping the bellows. The Examiner will be watching closely as it is unacceptable for the smoker to go out at the beginning of the hive manipulation. If the smoker has been well lit it will continue to burn slowly by itself if left in the upright position and produce an adequate supply of smoke when the bellows are worked.

1.7.3 See section 1. 6.1 above on the use of the smoker during the manipulation. We cannot stress too strongly that there is an art using the smoker which is easy to learn. Do observe the Master Beekeepers that you come across while they are working their colonies; you will learn a lot.

1.7.4 The Examiner will still be watching what you do with your smoker when the manipulation is complete, the hive closed up and the smoker is no longer required. Plug the nozzle with some green grass or better still have a wooden plug that fits snugly in the nozzle tied to the smoker permanently with a piece of string. On no account should the hot ashes be emptied out of the smoker, as they could be a fire hazard.

1.7.5 It will be noted that we recommend parking the smoker between your knees during a manipulation. We have met the odd beginner who has not felt comfortable using this method and prefers a hook on the rear of the bellows to hang it on the side of the hive. This is OK if cover cloths are not being used. A final point on the smoker is to ensure that it is in the possession of the manipulator all the time; we have seen many times someone 'borrowing' the smoker during a branch apiary meeting thereby leaving the manipulator without it when it is urgently needed.

1.8 Demonstrate the use of the hive tool.

1.8.1 Rather like the smoker there has been little written about hive tools and how they should be used; so perhaps we should start at square one on this particular topic. If the honeybee failed to collect propolis it would be a simple matter to take the hive apart and a hive tool would be unnecessary. Unfortunately all honeybees do collect propolis in greater or lesser quantities and one of its uses is to seal all gaps and crevices, making the dismantling process difficult for the beekeeper who has to resort to some kind of lever.

1.8.2 Let us first consider what jobs are required to be undertaken with this specially designed lever which has been given the name - hive tool. They are, as a matter of routine:

  a) To lever apart the supers and brood chambers of a hive.
  b) To lever up the crown board and scrape it clean after removal.
  c) To lever up the queen excluder.
  d) To split the bottom brood box from the floor.
  e) To remove the entrance block.
  f) To remove the dummy board and scrape it clean.
  g) To remove the frames from the brood chamber and supers (mainly horizontal leverage required).
  h) To scrape propolis and wax from top bars before reassembling the hive.
  i) To lever up the frames and dummy board in the brood chamber after an inspection.
  j) To lever up the frames in the supers before closing down the hive if they have been inspected.
  k) To clean out the remains of the ashes in the smoker before lighting.

The list of jobs is quite impressive but there are other uses which will become apparent after keeping bees for a few years, eg. removing drawing pins from frames so marked, scraping floor boards, etc. Most, if not all, hive tools have a 'gizmo' for prising up and removing nails, something we have never used.

1.8.3 The essential features of the hive tool are as follows:

  a) It should have a wide blade in order not to damage the woodwork of the hive. Red cedar is a soft light wood and is easily damaged. Full supers and brood boxes can weigh between 30 and 50 pounds and this load needs to be spread over as large an area as possible hence the wide blade (screw drivers and chisels are not good substitutes for a hive tool because of the narrow blade end).
  b) Because of the high loading the longest practical length is desirable thus needing the minimum force (or muscle power) to open up the boxes. Lengths vary between 6 to 10 inches (150 to 250mm) approximately.
  c) It requires a smaller lever at the other end of the blade to lever frames apart in a horizontal direction. Two versions are generally available, one in the form of a 'J' and the other formed into a semi-circle about ½ inch (12mm) diameter in the plane of the blade.

d) The tool should be made of good quality steel, preferably stainless, and should be rigid enough not to bend under normal use.

e) The end of the blade needs to be as thin as possible to insert between the boxes to be parted.

A typical appliance catalogue shows the following:

Lightweight J-type, heavy duty J-type, broad scraper type and stainless steel scraper ranging in price from £7.25 to £10.95.

The prices are quite unreal. Talk to your local builder and obtain a stainless steel building tie (cost c.50p); cut off the fish tail split end and grind a suitable blade to suit your requirements. We have two or three which we made some years ago; they work very well although somewhat short in length.

1.8.4 In the examination the Examiner will be watching you use the hive tool and there is a right and a wrong way to use it. Firstly it must become second nature to you to have the hive tool in your hand all the time; it is best held in the palm of your hand and held with the little and third finger while it is not in use. This will allow the use of the thumb and first finger to hold the frames as they are removed for inspection. Like the smoker it should not leave your possession during a manipulation. If you do want both hands and all fingers for some particular operation, temporarily put down the smoker and lay the hive tool across the top of the bellows. Pick up both the smoker and the hive tool when resuming the manipulation.

1.8.5 To remove parts of the hive and splitting boxes for removal, insert the blade of the hive tool between the two parts at one corner. To do this move the hive tool from side to side as it is pushed in; do not try to move the hive tool up and down it will damage the woodwork of the hive. We have seen this needless destruction many times at apiary demonstrations. Once the hive tool is in between the boxes gently lever in a vertical direction endeavouring to open the boxes without jarring. Once open they are easily held about an inch (25mm) apart to puff a little smoke between them before lifting off to the rear of the hive.

1.8.6 The other end of the hive tool now comes into play to remove the frames. First is to lever the dummy board away from the first frame towards the hive wall in a horizontal plane; this is best done at both ends and then lifted out. Do not hold the hive tool like pen or pencil between thumb and forefinger; hold it with a good grip and towards the flat blade to get the best leverage. This will allow a smooth parting of the sealed parts without jarring the frame or the dummy board. You now have room to work. Similarly, lever both ends of the top bar towards you to break the propolis and possibly wax seals and lift out vertically without rolling the bees on the adjacent frame. With practice you will find that once one side is moved with the hive tool it is usually possible to shift the other side by hand; this makes for a much faster inspection. Finally replace the dummy board at the opposite end of the brood box and use the hive tool to lever the frames hard up to one end by inserting the hive tool between the dummy board and the hive wall and levering hard. If you have never done this you will be surprised how much the frames can be closed up.

1.8.7 After use, scrape the hive tool clean to remove any wax and propolis. Wipe with a damp cloth to remove any stickiness.

1.8.8 Finally, there is an art in using the hive tool, just like the smoker. Watch carefully a Master Beekeeper at work on one of his hives, you will learn a lot. Watch carefully at a branch apiary meeting and you will more than likely notice some of the bad points of using the hive tool.

1.9 Remove combs from the hive and identify worker, drone and queen cells or cups if present, and to comment on the state of the combs.

1.9.1 During the manipulative part of the examination the examiner is likely to ask you to identify worker and drone cells. The difference in size is readily apparent even to the untrained eye. Note that there are approximately 4 drone cells per inch (c.6mm/cell) and 5 worker cells per inch (c.5mm/cell); both cells, if they are for brood rearing purposes, are approximately $^7/_{16}$ inches (11mm) in depth. If the cells are located in the supers then the depth of the cells can be considerably greater than those used for brood rearing. All drone and worker cells slope down towards the centre of the comb (septum) by a few degrees below the horizontal.

1.9.2 Queen cups are built by all colonies during the active season; in some parts of the country they are called play cells. They are reminiscent of the cup of an acorn and about the same size. They are always built in a position whereby they can be extended vertically downwards to form a queen cell. Normally they have a mat finish on the inside surface until they are polished ready for the queen to lay in them. The number built varies (say 3 or 4 to 20) from colony to colony and on the strain of the bee. When extended vertically downwards into queen cells the finished queen cell is about $1^1/_4$ inches (32mm) long and about $^1/_2$ inch (12mm) outside diameter with a pitted appearance on the outside rather like the case of a peanut.

1.9.3 We provide here a list of some of the more important items about combs that an examiner may ask:

1. Newly built to very old with the appropriate colour change from white to nearly black due to being varnished with propolis by the bees.
2. The ratio of the number of drone cells to worker cells. It is generally accepted that only very few drone cells should occur on a brood comb. A healthy well balanced colony will have 1% of its population as drones, the other 99% will be workers.
3. Irregular comb with holes through the comb from one side to the other.
4. Pollen-clogged comb. If the comb becomes full of pollen it acts as a barrier in the brood nest for the orderly laying of the queen. These combs are usually found in the spring with stale pollen from the previous autumn.
5. Comb containing granulated honey. The situation here is similar to the pollen-clogged comb and is quite common in areas where the colony has been working rape the previous season. Both types should be removed from the brood chamber.
6. Comb contained in damaged frames. In most cases the damage occurs to one of the lugs due to careless use of the hive tool or the rotting of the wood due to unsuitable fixing pins corroding.
7. Comb contained in frames with 'sick' fastenings (ie. where the nails or gimp pins have rusted). The wood around the rusty nails becomes rotten and the fastenings then are unsound.
8. Comb that does not fully fill the containing frames. These are generally due to the foundation having not been drawn out properly due to an inadequate flow or a weak stock.

9. Comb that is not symmetrical in the frame. This can be due to the foundation not having been fitted securely and the bees drawing the foundation badly.

10. Comb that is built on either wired or unwired foundation. In the brood chamber with old combs it is difficult to determine whether the foundation was wired whereas if the frame is wired it is easy. Supers with translucent white wax can be held up to the light to reveal the wiring.

11. Comb attached to the adjacent comb before removal. This sometimes happens in supers with a wider frame spacing than the brood chamber when the colony is weaker than it should be or the flow is indifferent when the foundation is being drawn.

12. Brace comb (horizontal) is usually due to narrow top bars ie. $^7/_8$ inch (22mm) or bad spacing of frames. Burr comb (burr=any impediment or inconvenient adherent) may be found on the edges of frames. Both types are best removed with the sharp end of the hive tool or a sharp knife.

1.10 Identify the female castes and the drone.

It should be noted that the syllabus recognises two female castes and the drone. We believe that the question of two castes or three is a matter of opinion, there being arguments in favour of both. As three have been recognised for over a century and three are used in most of the literature the change to two from three seems unnecessary and confusing. Therefore, we still like to use the three castes of the honeybee namely, the queen, the worker and the drone. In order to identify the three castes we provide a detailed description of the external structure of each.

The honeybee developed from a worm-like ancestor 30 to 40 million years ago. This worm-like animal is common to all the insects and had 19 segments. In the case of the honeybee the first six developed into the head, the next four into the thorax and the last nine into the abdomen. Close observation of a larva will clearly indicate the head and 13 other segments; the 6 segments making up the head are not distinguishable. In the adult honeybee 4 segments can be seen to make up the thorax and 6 segments making up the abdomen. If 6 segments are contained in the head where is the balance of 3 to make 19 in total? The answer to this riddle is that the last 3 are contained internally to form the sting chamber.

1.10.1 Structure common to all three: head, thorax and abdomen:

- Head: 2 compound eyes, 3 simple eyes (ocelli),
  2 antennae (scape + flagellum),
  2 mandibles,
  1 proboscis.

- Thorax: 4 segments (T1 to T3 plus A1),
  2 pairs of wings,
  3 pairs of legs (on segments T1, T2 and T3).

- Abdomen: 6 visible segments (A2 to A7),
  3 invisible segments A8 to A10 (which are internal and part of the sting chamber).

1.10.2 Physical size:

Worker about $^5/_8$in (16mm) long.

Queen about 1in (25mm) long but larger than a worker in diameter; nb. queen excluder.

Drone about $^3/_4$in (19mm) long (much fatter than q. or w.).

1.10.3 Head:

Worker - triangular in shape with long proboscis.

Queen - similar to worker but rounder with short proboscis.

Drone - almost circular (nb. large compound eyes). Antenna has extra joint (12 segments) and the mandibles are very small. The proboscis is short.

1.10.4 Thorax:

Worker/queen - similar in size; dorsal side in the queen appears hairless cf. a worker.

Drone - larger/stronger, larger wings (stronger flier).

1.10.5 Abdomen:

Queen - very distinctive (long/tapering).

Drone - also distinctive (fat and furry).

Worker - specialised (wax glands and Nasonov gland).

All castes have 10 pairs of spiracles (entrance to the breathing tubes of the honeybee) on segments T2 to A8, the last being invisible and inside the sting chamber.

1.10.6 Legs:

All have the same formation - 4 sections, 5 joints and a foot (tarsus with 5 subdivisions).

Note: the forelegs of all 3 castes have an antenna cleaner and only workers have pollen-collecting equipment on the rear legs. The hairs on the forelegs are used as a brush for cleaning the eyes and head in all castes.

1.10.7 Wings:

The four wings are similar in all three castes. The forward pair are large with a fold to engage with the hooks on the smaller rear wing. Drone wings are much larger. In all 3 castes the wings are folded at rest and lie flat on the dorsal side of the abdomen.

1.10.8 Hair:

The whole of the exoskeleton is covered in plumose hairs which have an important function in the worker for trapping pollen before the bee transfers it to the pollen baskets on the hind legs

During the examination the Examiner is likely to ask you to identify the three castes during the manipulation at the hive and perhaps ask some simple supplementary questions based on the above descriptions of the external aspects of the castes concerned. One of the best ways to become familiar with the anatomy of the honeybee is to take live specimens from a hive of bees. Pick the bees up by the wings, put them in a match box and place them in the freezer for a couple of hours. On removing the sample from the freezer allow it to thaw out before examining the external structure of the bee with a magnifying glass at your leisure. Just remember not to take the queen when choosing your sample.

1.11 Identify brood at all stages.

1.11.1 The definition of brood means all those forms of life excluding the adult form (or imago) before the honeybee emerges from its cell. This includes the egg, the larva, the pro-pupa and finally the pupa which has the adult form.

1.11.2 The egg. Good eyesight and a little practice is required to identify eggs in worker cells which are easier to see in good lighting conditions or bright sunlight. The egg is pearly white in colour and approximately 1.5mm long and about 0.33mm in diameter being slightly larger at one end (eventually the head of the larva). It weighs 0.13mg. When the egg is first laid it is stuck, end on, to the base of the cell and during the next 3 days it moves through 90° until it is resting on the bottom of the cell just before hatching. If you have difficulty seeing eggs then ask a Master Beekeeper to help you; he will more than likely shake all the bees off the frame and widen the mouth of a cell containing an egg with a ball point pen for you to see. It is essential for the examination to be able to identify eggs in worker cells.

1.11.3 The larva. The larva always adopts one position in its cell, that is, curled up on one side laid in a pool of brood food or 'bee milk'. As it is fed progressively it rotates in its cell, rotating as it eats the food. The larva can only breathe through the ten breathing inlets (the spiracles) on the side which is not in the brood food. Healthy larvae are, like the egg, pearly white in colour and always shiny. The larva, when it hatches, is about the size of the egg and after 5 days it completely fills the cell in a curled up position, its tail touching its mouth. Just before the cell is sealed it weighs about 114mg; an incredible change in weight from 0.13mg in 5 days, an increase of about 850 times. Identifying the age of the larva is somewhat subjective but for practical purposes the difference in size from the egg to the 5 day old larva can be divided into 5 equal increments. It is unlikely that an examiner would fail a candidate if a minor error occurred on the 'guesstimate' of the age of the larva.

1.11.4 Sealed brood. This is the easiest stage to identify. Healthy brood is capped in a mixture of wax and pollen to allow the larva and pupa to breathe. The cappings are convex, coffee-coloured with a dry matt finish. Sealed brood (see next section 1.12) is not to be confused with sealed pollen or old capped honey which are capped with pure beeswax which has a slightly shiny finish.

All the above comments are applicable to both worker and drone brood. Queen cells developed from queen cups are quite different and readily identifiable.

1.12 Demonstrate the difference between drone, worker and honey cappings.

We believe that the heading for this part of the syllabus should read "Demonstrate the difference between drone and worker brood cappings and honey cappings" because honey can be stored in either worker or drone cells and capped when full and ripe. There is one other capping that the Basic student should know and that is the capping on pollen stored for the winter period, often known as pickled pollen.

1.12.1 Drone and worker cells are easily distinguished one from another by their physical size. Both are, of course, hexagonal with an angle of the hexagon uppermost and lowermost and the two sides of the cell having parallel sides. There are approximately 4 drone cells to the inch and 5 worker cells to the inch, easily distinguished by eye. In area, there are 25 worker cells to the square inch and 16 drone cells approximately.

1.12.2 Cells that contain brood are capped with a mixture of wax and pollen. The capping is convex when viewed from the outside of the cell. This mixture of pollen and wax is usually a coffee colour and provides the necessary porosity for the larva to breath during its confinement in the cell. It also encourages the adult bee (or imago) to chew away the capping when emergence takes place as pollen is part of the adult bee's natural diet. These coffee coloured cappings have a matt finish to them which is quite different from the wax-like finish on sealed honey.

1.12.3 Honey cappings are also convex in shape but are made from pure beeswax and are initially white when capping is newly completed. Some strains of bee leave a small air space below the capping and these cappings look very white while other strains of bee allow the honey content of the cell to touch the underside of the capping and thereby giving it a darkened appearance; the darker the honey the darker will be the appearance of the capping. With time the cappings of honey in the brood chamber tend to darken; this is particularly noticeable in the spring where the bees have not used some of their winter stores. The importance of using wax is to completely seal the honey from the damp air in the hive. Honey is hygroscopic and readily absorbs water. If water is absorbed then the honey is diluted to such an extent that it would start to ferment in the cells and become useless as stores for the honeybee. When fermentation takes place then alcohol is produced and this in turn causes dysentery in the bees.

1.12.4 Finally there is 'pickled pollen'. Pollen collected in the autumn for use in the late winter and very early spring is packed into the cells around the brood chamber and topped off with a thin layer of ripe honey before being capped with pure beeswax. The cappings are virtually identical with honey cappings but nearly always appear quite dark in colour.

During the course of the manipulation part of the Basic Examination the Examiner will point out various cells for the Candidate to identify.

1.13 Identify stored nectar, honey and pollen.

1.13.1 Nectar is an aqueous solution of various sugars. Other substances such as vitamins, enzymes, trace elements, organic acids, nitrogen compounds and aromatic substances are present in small amounts. Nectar varies from plant to plant due the type of the soil supplying the nutrients to the plant and the type of species or sub-species of the plant. Other variations may be due to the position of the blossom on the plant or tree and the temperature of the environment. Nectar is collected by the foraging bees and transported back to the hive where it is received by house bees and elaborated by the bees into honey. During this transformation the nectar is placed in the cells of the hive and water content is reduced by the bees fanning at the entrance of the hive, promoting currents of air within the hive, whilst moist air is extracted at the entrance. Nectar will be found in the hive:

• In the active season when the brood nest is inspected for signs of swarming and colony health, nectar which is not yet changed into honey is very aqueous and will fall out of the cells as the frames are manipulated by the beekeeper. This is called 'unripe honey,' it can be seen as a glistening colourless liquid in odd cells in the brood nest.
• Nectar/unripe honey is easily identifiable during a 'flow' when it seems to fill every empty cell in the brood nest. In these conditions the open cells in the combs of the super frames are also full of nectar/unripe honey. If one of these super frames with open cells containing nectar is held horizontally and given a sharp shake, nectar which has not yet been elaborated into honey and sealed in the cells with a wax capping, will fall onto the top bars of the frames below. These drops of nectar are referred to as 'unripe honey' ie. the water content is still too high. It is not until the bees have reduced the water content to less than 20% by evaporation that the nectar is completely transformed into honey. Unripe honey should never be extracted by the beekeeper as the high water content will cause the extracted honey to ferment. Fermentation will also occur in the open storage cells of the hive if sugar syrup for winter stores is fed late in the year ie. when the daily temperature drops at the end of the summer accompanied by cold moist air and shorter daylight hours. In weak colonies the bees are unable to reduce the water content and seal the stores with wax. Fermenting stores when ingested by the bees cause dysentery due to the alcohol content.
• See section 3.7. which details how nectar is transformed into honey.
• Early in the year, late February or March, there is scarcely any nectar for the bees to collect, no nectar is to be seen in the cells of the combs and the bees are consuming the stores of honey prepared the previous autumn.

1.13.2 Honey is nectar which has been transformed by the honeybees and stored in sealed cells above the brood nest to be used in times of dearth. The water content of the nectar is reduced to c. 20% or less and the sugars have been inverted during the process of elaboration by the honeybees. The cells when full are sealed with a capping of wax. This preserves the honey until required by the colony. The cappings vary in colour and shape depending on the type of nectar being collected, the position of the honey cells, the age of the stored honey and the expertise of the strain of the bees in the hive in sealing the cells. See section 1.12.3. In the wild, bees prefer to store honey in drone comb, which is a more economical use of wax. Foundation supplied by the beekeeper in the honey supers may be impressed with either drone or worker hexagonal impressions; sometimes starter strips are fixed below the top bars depending on the preference of the beekeeper. When learning to recognise sealed honey stores in the arches above the brood nest in the brood box and to differentiate between sealed pollen, sealed brood and sealed honey try testing your skill by uncapping a cell with the tip of the hive tool or a sharp knife and confirm your assumption as to the contents.

1.13.3 Pollen is the male germ seed of plants. It is carried back to the hive in the *corbiculae*, pollen baskets, one on each of the outer surfaces of the rear legs of the worker bee. Pollen varies in colour though shades of yellow predominate, the original colour varying as it is preserved by the honeybees. The pollen colour can be different from the outer skin of the anthers which contain the pollen grains. This pollen may be for immediate use or may be stored in cells:

- directly above the brood nest and below the arch of honey. Here it is easily reached by the nurse bees tending the brood.
- in an extended arch in the first super above the brood nest.
- in odd cells in the super frames
- in the cells of frames on the outer side of the brood nest.

When the bees of a hive are hard pressed to cope with the influx of nectar and pollen and there are no prepared empty cells around the brood nest, then the pollen is deposited in any available cell by the foraging bee on her return to the hive. The foraging bee seeks out a suitable cell and deposits the pellets of pollen in an empty or partly filled cell in the vicinity of the brood, this pollen has already been mixed with honey and a fugitive substance which will prevent the nucleus of the pollen grains developing to maturity. Later a 'house' bee will pack the pollen into the cell with her forelegs at the same time mixing honey with the layers of pollen, see 1.12.4. Sealed cells of pollen can be mistaken for sealed honey stores.

A super frame of sealed honey if held up to the light should appear translucent. Granulated honey or pockets of stored pollen, especially the black pollen of the poppy flower, *Papaver rhoeas*, will show as dark dense cells. Pollen and or granulated honey in the super frames will give problems when the honey is extracted. The frames are also spoilt for entering into Honey Shows.

Any pollen which has not been pickled with honey and sealed with wax for use in times of dearth mainly, ie. October to February, will eventually become infected with fungus, *Bettsia alvei,* or the pollen mite, *Carpoglyphus lactis*. The invading fungus turns the pollen white with a hard consistency. The pollen mite frequently found on frames of pollen stored outside the hive cause the pollen to disintegrate into a light brown powder. Both infestations render the pollen useless to bees as a source of protein.

1.14 Take a sample of worker bees and put them in a match box or similar container.

1.14.1 It is quite remarkable the number of candidates who come forward for the basic examination and reveal that they have never picked up a bee or collected a sample for adult bee disease analysis and often have no idea how to proceed. There are two points worthy of note which are as follows:

1) It is to be noted that it is virtually impossible to pick up a bee without damaging it when wearing gloves.
2) A cage (eg. a Butler cage) is unsuitable for collecting a sample of honeybees for adult bee disease diagnosis because 30 bees are required.

1.14.2 Putting 5 or 6 bees in a Butler cage is easily done by picking them up by the wings and facing them into the open end of the cage and releasing them; cover the end of the cage with your thumb as soon as each bee has entered the cage. Select those bees which are gorging themselves with honey after being smoked. If the bees are intended to accompany a queen in transit young bees capable of producing royal jelly should be chosen.

1.14.3 There are a variety of ways of collecting a sample for adult bee disease diagnosis:

• Perhaps the easiest and quickest method is with a matchbox. Have it open in one hand, holding a frame of bees in the other and slowly manoeuvre the open matchbox over the bees and wriggle it slowly shut. If there are plenty of bees on the frame, it will hold just about the required 30 bees for the sample. Old bees are required for disease diagnosis so always take the bees from the end frames of the brood chamber.

• Another method of collecting outgoing foragers, which are the older bees, is to put a large clear plastic bag with its open end formed into a rectangle about 6 inches × 18 inches (150mm x 450mm) with a wire frame and count 30 outgoing foragers flying into the bag. The difficulty with this method is killing them for analysis if this is to be done immediately or getting them into a suitable sized box for sending through the post to the microscopist. We do not recommend this method.

• Covering the entrance with an old sack or blanket for a few minutes when the bees are flying well will very quickly have enough returning foragers on the cover to pick up 30 individually. Again we do not recommend this method; it takes too long.

• Finally, it has been recommended elsewhere that the easiest way is to put a honey jar over the feed hole (assuming there are no supers on the colony) and collect a sample which rises from the brood chamber into the jar. We are of the opinion that there is a risk of collecting young bees instead of old bees and on these grounds consider it an ill-advised method.

1.14.4 Most beginners and those attempting the Basic Examination are likely to be sending their samples of bees to their Branch or County Microscopist. If so, then the following points should be observed:

a) First check that your microscopist can accept your sample(s) and that he is not away on holiday or indisposed in any way.

b) Take the sample and kill the sample immediately by putting the matchbox full of bees in a small plastic bag and then placing the bag with the box of bees in the freezer compartment of a refrigerator where the temperature is about -15° to -18°C.

c) The following morning send them by post to the microscopist ensuring that each sample is identified with the apiary name and the hive number appropriate to each sample. Remember that decomposition of the sample will start as soon as it is removed from cold storage and the microscopist wants the sample to be as fresh as possible for dissection purposes. Remove the sample from the plastic bag and place in a stout paper envelope; plastic packaging should not be used as it makes the sample of bees moist. First class postage is required, going into the post on Monday to Friday only. Don't post them on a Saturday or a Sunday; keep them safe and sound in the freezer until Monday morning.

d) Enclose a stamped and addressed envelope for the microscopist to send you his written report and any recommendations for treatment. Some of the better microscopists telephone the results in addition to the mandatory written report.

1.14.5 The syllabus doesn't include the time of the year when samples should be taken. We regard this as very important. On a regular basis every beekeeper should sample each hive twice per year. The first time is in the spring at the first inspection and the last time is in the autumn when supers have been removed and before winter feeding is commenced. The spring tests are to determine whether the colony has disease or whether the queen is starting to fail if normal spring build up of the colony does not occur. In the autumn it essential that the colonies with adult bee diseases are treated before clustering commences for winter, the time when diseases are spread more readily from one bee to another.

1.14.6 If piles of dead bees are found outside the hive and poisoning is suspected then a very much larger sample of bees is required; at least 3000 bees. Collect all the dead bees and divide the sample into two. Put one half into a plastic box inside the freezer and send the other half in a stout cardboard box to National Bee Unit (NBU), Sand Hutton, York, North Yorkshire Y04 1LZ giving details of the incident. First phone the NBU for advice and contact the Secretary of your local Association; other beekeepers may be involved if the poisoning is due to the spraying of crops with pesticides. The bees in your freezer are the final bit of evidence, if the other sample sent away for testing is lost before analysis has been completed.

1.15 State the number of worker bees required for an adult disease diagnosis sample.

There is no particular number of bees required for an adult bee disease sample. However, it is generally agreed that about 30 is a satisfactory sized sample. The larger the sample the more accurate will be the assessment of colony infection and conversely the smaller the sample the less accurate will be the assessment. The probability curves have a knee in them occurring at about the 30 bees point; there is very little change in the curves for greater sample sizes, so hence the 30 figure If you have only say 25 in the sample do not worry, the microscopist will count them and provide the correct answer for that sample size.

1.16 Demonstrate how to shake bees from a comb and how to look for signs of brood disease.

There are two diseases in particular, American Foul Brood (AFB) and European Foul Brood (EFB), where it is essential to shake the bees from the combs to look for signs of these diseases. Both diseases are diseases of the brood and it is important to note that there are no signs of either disease associated with the adult bees in an infected colony. In order to diagnose either disease in the field, it is necessary to open up the colony and examine the combs containing brood where the signs are to be found.

To do this properly it is necessary to shake the bees off the comb before examining it, leaving no more than a few bees on the comb. The reason for this is that in the early stages only an odd cell or two will be exhibiting the tell-tale signs of disease. This important aspect of searching for the diseases is frequently overlooked and inadequately expressed in much of the literature.

There is a right and a wrong way of shaking bees off combs, the objective is to rid the comb of bees and keep them in the hive (not flying around the apiary); therefore after removing one frame or dummy board from the brood box to allow sufficient space, raise a frame of brood and bees about 2 inches (50mm) above the top bars of the surrounding frames holding the frame firmly by the lugs, shake it sharply in a vertical direction inside the brood chamber without jarring the rest of the colony. The bees removed from the comb will fall inside the hive on to the floor. This will keep most of the bees in the hive and if they are in the hive they are under the control of the beekeeper. It is important that all beekeepers learn to undertake this procedure early in their beekeeping career and to be able to go through the whole colony whilst keeping it under control.

If you are in any doubt about this manipulation then approach a Master Beekeeper and ask him/her to demonstrate the technique for you.

** ** ** **

2.0 EQUIPMENT

2.1 Name the principal parts of a modern beehive.

2.1.1 In order to make this section more realistic we provide below a list of hives that can be found in the UK, broken down into single and double walled types:

- Double walled hives:

(a) WBC	named after William Broughton Carr	B *
(b) Burgess Perfection	not now produced commercially	B

- Single walled hives:

(c) British National	first attempt to standardise	B
(d) Modified National	introduced 1960	B *
(e) Smith	developed in Scotland for BS frame	T *
(f) Modified Commercial	attributed to Simmins	B *
(g) Langstroth	most widely used hive in world	T
(h) Dadant	first developed in USA by Dadant	T
(i) Buckfast Dadant	developed by Bro. Adam	B

- Other types:

(j) British deep, Catenary, Long hive, Cottager, Conqueror, etc. together with the skep woven in straw in a variety of shapes and sizes. Copies of some of the standard single walled hives are being made in high density polystyrene.

* these hives may be regarded as the most popular in use in the UK; the Modified National being used in the greatest numbers.
B or T = bottom or top bee space. It should be noted that all the single walled hives marked * may be constructed with either top or bottom bee space. The two systems cannot be mixed.

It will be clear from the above that for the beginner the array of different types of hive is bewildering.

2.1.2 The essential parts of all the hives are:

• Floorboard and entrance block: the double walled hives (a) & (b) have infinitely variable slides in lieu of an entrance block. The roof should be capable of storing the entrance block.

• Brood chamber: (a), (c), (d) & (e) are all designed for the British Standard (BS) frame with long ($1\frac{1}{2}$ inch or 38mm) lugs except (e) which has short lugs. The other brood chambers are characterised by their different sizes, all using larger frames than the BS. The outer dimensions of

24

(d) & (f) are very similar and can be used together particularly the supers of each. Similar to the Smith hive, the Commercial hive uses frames with short lugs, thereby simplifying its construction.

• Supers: the supers for all hives are shallower than the brood boxes. In the case of the Langstroth hive it is common to use the same size boxes throughout, even though shallow supers are available. This has advantages of standardisation but the supers are very much heavier when full, demanding a greater muscle power. Supers for most hives can be equipped with a variety of frames (see section 2.5).

• Crown boards: are usually provided with one or two holes which can be used for ventilation, feeding or, fitted with Porter bee escapes, may be used as a clearer board. Depending on the type of hive, a bee space will be provided on the underside. They are generally made of plywood, but can be panelled in glass (a bit of a gimmick for serious beekeeping) which causes condensation on the underside.

• Roofs: these come up in a variety of designs of varying depths from about 3 inches (75mm) in the Langstroth to 6 or 9 inches (150 or 230mm) in other hives. Generally, a deep roof with a minimum clearance between the inside dimensions and the outside of the hive is best in order to prevent it being blown off in windy conditions. All roofs have ventilators, considered by many serious beekeepers to be generally inadequate in size. The standard designs provide wire mesh on the inside allowing them to become blocked from the outside by solitary bees, etc. Putting the mesh on the outside prevents this happening.

• Dummy board: an essential piece of equipment which should be found in every brood chamber and should be considered an integral part of every hive. In our opinion, it is wrong to fill a brood chamber with frames and no dummy board; there should be a dummy board and enough room on its outside face to insert a hive tool to lever the frames tight up together. When the dummy is removed there should be enough room to manipulate the colony without having to remove one end frame.

2.2 The concept of the bee space and its significance in the modern beehive.

2.2.1 Definition and description of the concept of "bee space".

The Rev. L.L.Langstroth of Philadelphia USA is credited with 'inventing' the bee space in 1851/2. He showed that with a bee space of ½ inch (12mm) between parts of the inside of the hive, the bees would make little attempt to construct brace and burr comb. He found that by observing this bee space, parts of the hive could be made moveable and interchangeable. This was the turning point from skep beekeeping to modern day beekeeping with the moveable frame hive. The salient points relating to bee space in the moveable frame hive are as follows:

a) Bees will propolise a space less than $1/4$ inch (6mm) and will build brace or burr comb in a space greater than $3/8$ inch (9mm).

b) Bee space is now considered to be $^5/_{16}$ inch (8mm) thereby allowing $^1/_{16}$ inch (1.5mm) tolerance above and below this figure to cater for expansion and contraction of the woodwork while the parts are in use.

c) The bee space in a modern hive includes the space between boxes of frames and between the frames and the crown board, the space between the wall of the hive and the side bars of the frames, the space between the walls and the end combs and lastly the space between adjacent top bars and side bars of the frames. All these should be $^5/_{16}$ inch (8mm).

d) There is one exception in most hives and that is between the bottoms of the frames in the brood chamber and the floorboard which is of the order of 1 inch (25mm). The reasons for this are a bit obscure; in practice the authors have found that during rapid colony build up in the spring much drone comb is built in this area by extending downwards the comb in the brood frames. It also provides a parking space for bees in a large colony in bad weather.

e) If a frame spacing of $1^3/_8$ inch (35mm) is used the space between adjacent comb faces becomes $^1/_2$ inch (12mm), ie. two bee spaces of $^1/_4$ inch (6mm) allowing the bees to work the two comb faces back to back. If the frame spacing is $1^1/_2$ inch (38mm) then the inter-comb space increases to $^5/_8$ inch (16mm), ie. two bee spaces of $^5/_{16}$ inch (8mm).

f) In the brood chamber the only combination of frame and frame spacing dimensions that fully meets the bee space criterion is frames with top and side bars = $1^1/_{16}$ inch (27mm) wide with $1^3/_8$ inch (35mm) spacing between frames.

g) In supers with $^7/_8$ inch (22mm) frames and $1^7/_8$ inch (48mm) spacing between frames, the space becomes one inch (25mm) between adjacent frames which is very much greater than a bee space.

2.2.2 The significance of the bee space allowed the construction of moveable frame hives whereby the frames can be easily and readily removed for inspection of the colony. This discovery revolutionised beekeeping management and swarm control. All the developments in both equipment and management techniques stem from the discovery of 'bee space' and have occurred since 1851.

2.3 Assemble a frame and fit it with wax foundation.

This is a very easy part of the syllabus but, like all things, there is a right and a wrong way of doing it. It is best that it is considered in two stages, one, the frame assembly and two, fitting it with foundation afterwards.

2.3.1 Frame assembly.

All frames consist of 5 parts namely, the top bar, two side bars and two bottom bars. The side bars usually have a slot on the inside surface which is used as a location for the sheet foundation. The top bar comes from the manufacturer with its wedge still attached to the top bar by a sliver of wood; this needs to be removed by cutting it away with a Stanley knife and cleaning out the angle of the wedge in the top bar. Clean up the wedge as well so that its cross section is a clean rectangle. It is wise to assemble one without pinning any of the joints; you are likely to find that the joints between the side bars and the top bar are very tight, often too tight and sometimes the side bars crack or in the worst case split lengthwise. They are deliberately made tight to be assembled without glue. If the joints are too tight the top bar joints can be eased slightly with a fine file. Ensure that all the joints have fully mated (lightly tap them home with a hammer) and then check that the adjacent bars are at right angles one to another. The last check before pinning is to ensure that the frame is not skewed (ie. it should lie in one plane) by looking at it sideways and ensuring that both the side bars are parallel to each other. Alternatively, check that the top bar is parallel to the bottom bars. If a frame is skewed then the comb will be drawn out in an irregular fashion when used in the brood chamber or super.

The joints can now be pinned with blued $^3/_4$ inch (19mm) long gimp pins. They are best stored in a pot and adding a few drops of oil; replace the lid and give it a good shake to get a thin film of oil on each pin. Each joint between the top bar and the side bar needs two pins inserted and tapped home so that they are at right angles to the grain of both the top bar and side bar (ie. pinned from the sides of the side bars). Hoffman self spacing frames with a 'Vee' on one side of the side bar shoulder and a flat on the opposite side still need two pins; very often inexperienced beekeepers pin them only on the flat side. Pinning the bottom bars entails two pins into each side bar but his time the pins enter the side bars parallel to the grain in order that they can be removed easily for rewaxing the frame. Some prefer to assemble the frame with one bottom bar missing to make it easier to insert the wax foundation; this is a matter of personal preference. Finally, after pinning, check again that the frame is square and without skew.

The above will see you through your examination but for your own use we recommend that all frames are assembled with their joints glued and in lieu of the steel gimp pins, $^3/_4$ inch (19mm) long copper nails of thickness $^1/_{16}$ inch (1.5mm) are used. The side bars and the bottom bars will require pre-drilling for these copper nails. Frames put together this way will last a lifetime; frames not glued and assembled with steel gimp pins will develop nail sickness in a few years despite the blueing and pre-oiling which just puts off the inevitable.

2.3.2 Fitting a frame with foundation.

The wax foundation used for frames can be either wired or unwired together with an alternative arrangement whereby the frame is wired and the foundation is then fitted to the frame and the wire at the same time. For this examination, only wired or unwired foundation would be required as it is a more difficult and time consuming job to prepare a wired frame (which ultimately provides a superior frame and comb).

The wire in wired foundation is usually these days in the form of a continuous zig-zag with small loops of the wire protruding from the upper and lower edges of the sheet. The first job is to very carefully bend the loops of wire along one edge only, at right angles to the plane of the sheet of foundation without breaking the wax at the point where the wire is bent. The sheet is then installed in the frame by sliding it along the slots in the side bars until the bent wires fit snugly in the rebate for the wedge bar. Clearly this is easier if one bottom bar is missing at this stage. The wedge is then placed in position and pushed up into the angle of the wedge thereby clamping the bent wires and the edge of the wax sheet. Next the wedge is pinned using the same sort of gimp pins used for assembling a frame.

Three pins are required and put in the same direction as the pins fastening the side bars to the top bar (ie. at right angle to the wax sheet). No other way is acceptable because the point of the pins will protrude from the top of the top bar if they are pinned incorrectly; further, the pinning in this case does not exert any pressure on the wax sheet along its edge and therefore would allow the foundation to slip, particularly if it was unwired foundation. Finally the second bottom bar can be fitted and pinned into position. Again check for squareness and absence of skew.

The procedure using unwired foundation is identical, except that it can be fitted with both bottom bars in place. It is even more important with unwired foundation to get the pinning of the wedge bar in the correct direction.

2.4 The reasons for the use of wax foundation.

2.4.1 The purpose of wax foundation is to induce the bees to build straight comb in wooden frames thereby allowing:

> a) easy inspection of both sides of the comb and facilitating inspection of every cell on the comb face,
> b) easy extraction of honey,
> c) easy manipulation of the colony,
> d) re-use of the wooden frames,
> e) either worker or drone comb to be constructed as required by the beekeeper,
> f) minimising the amount of wax that has to be produced by the bees (the bees use the extra wax contained in the extra thickness of foundation compared with natural comb).

2.4.2 The types of wax foundation available are as follows:

> a) thick foundation, wired in brood chambers and wired or unwired for use in supers,
> b) thin foundation to be used for cut comb, sections or Cabanas,
> c) all foundation may be embossed with either worker or drone cells,
> d) sheets are available to fit most types of frame for most types of hive.

2.4.3 The wired foundation is in different styles depending on the manufacturer or supplier as follows:

> a) the wire can be either straight or crimped,
> b) the positioning of the wire in the sheet can be either horizontal or diagonal formation (the terminology is a bit misleading - actually a series of V's from top to bottom),
> c) the material varies from tinned iron, stainless steel to Monel metal.

2.4.4 Historical:
Mehring in Germany produced the first wax foundation in 1857, not long after Langstroth invented the first moveable comb hive. Weed in the USA produced the first machine to roll foundation in quantity on a commercial basis. Many beekeepers these days make their own foundation as a DIY activity using a simple press or a die between rollers. Foundation for BS

brood frames has about 8 sheets per lb of wax and for Commercial frames about 5 sheets or 18 sheets/kg and 11 sheets/kg respectively.

2.5 The spacing of the combs in the brood chamber and super for both foundation and drawn comb and methods used to achieve this spacing.

2.5.1 To become familiar with the various types of spacing it is essential to see and handle the bits of equipment involved. There are two major methods of spacing frames. Firstly, self spacing where the dimensions of the frames automatically provide the correct distance between the frames when they are placed in the hive. Secondly, frames can be spaced by attaching specially designed spacers to the frames. Both methods are widely used in the UK.

2.5.2 It is important when manipulating a colony to always ensure that the frames are levered up tight to one side of the brood chamber. Failure to do this each time, will result in a build up of propolis on the spacing surfaces and the spacing gradually becoming too large. Also it prevents, to some extent, uneven combs which are thick at the top with an arch of honey and are consequently difficult to move to a different position in the brood chamber (note that wide top bars and $1^3/_8$ inch (35mm) spacing obviates this trouble). Generally it is the beekeepers who do not use a dummy board that experience this problem; it often gets to the stage where the first frame has to be levered out and forcibly pushed back which is very difficult to do without damaging some of the bees.

2.5.3 Self spacing: frames are used in both the brood chamber and in supers. There is one type for brood, the Hoffman frame where the side bars have specially shaped 'shoulders' with a 'V' on one side and a flat on the other. The overall dimension between the 'V' and the flat is $1^3/_8$ inch (35mm) giving the same spacing between centres when the frames are placed in the hive. The self spacing frame for supers is the Manley frame which has parallel sided side bars $1^5/_8$ inch (41mm) wide. These provide $1^5/_8$ inch (41mm) spacing between the centre lines of the comb when placed in the hive.

2.5.4 Spacers for attachment to frames: are listed below with their spacing dimension.

Metal ends -	for brood frames, originally $1^9/_{20}$ inch (37mm) but now $1^7/_{16}$ inch (36.5mm), for super frames, $1^7/_8$ inch (48mm).
Plastic ends -	for brood frames, $1^7/_{16}$ inch (36.5mm), for super frames, 2 inch (50mm).
Double V plastic ends -	for brood frames (min. contact area), $1^7/_{16}$ inch (36.5mm).
Hoffman adapters -	for brood frames (2 types available), both $1^3/_8$ inch (35mm).
Yorkshire spacers -	for brood frames, approximately $1^1/_2$ inch (38mm).

2.5.5 Other spacing methods: include castellated metal strips for use in supers only and screws or studs on the side bar edges. We have found one or two cases where the beekeepers concerned use 'finger' spacing or spacing the frames by eye. This is not recommended except as a temporary measure (nb. it would be very foolish to move a colony without proper spacers).

2.5.6 Additional comment. After many years of beekeeping and because it is difficult to breed a bee which collects a minimum of propolis, all our frames and spacing have been modified. Our modifications make colony manipulations easy and pleasurable with virtually no hassle. We commend it to anyone starting out on their beekeeping career.

Brood frames:
a) $1^3/_8$ inch (35mm) spacing Hoffman with the 'Vee' planed off and replaced by a #6 stainless steel pan head screw $^3/_4$ inch (19mm) long adjusted to give the original $1^3/_8$ inch (35mm) spacing. Wide top bars, $1^1/_{16}$ inch (27mm) only should be used.
b) $^5/_{16}$ inch (8mm) should be cut off each end of the top bar and replaced by two #6 stainless steel pan head screws 1 inch (25mm) long.

Super frames:
a) Manley frames with $1^5/_8$ inch (41mm) spacing. Cut off $^7/_{32}$ inch (6mm) from the edge of each side bar and replace it with a #6 stainless steel pan head screw $^3/_4$ inch (19mm) long in a similar way that the brood frames were modified.

We have seen so many times beginners to beekeeping being sold equipment which is not in the best interests of beginners. The vendors are generally well meaning but sadly lacking in experience to advise. *Caveat emptor*!

** ** **

3.0 NATURAL HISTORY OF THE HONEYBEE

3.1 An elementary account of the production of queens, workers and drones in the honeybee colony.

The constitution and size of a normal colony varies during the course of a year starting off in the spring with a few thousand workers (winter bees) that have over-wintered with the queen. With the advent of spring and the flowers associated with it, nectar and pollen are collected, the queen starts laying more and more eggs per day and the colony increases in size. It reaches a peak of about 40,000 workers (summer bees with a life span of six weeks) at the end of June. The main honey flow starts in July and the colony then starts to decrease in size. In spring drones are produced in small numbers reaching a peak of about 400 (1% of the worker force) and they are evicted from the colony after the main honey flow when their mating function has been completed. During the spring to summer build up the colony may swarm and in order that this can be effective, new queens are produced, one of which will head up the old colony. Generally, the old queen leaves with the swarm to find a new nest site. The stores collected during the summer provision the colony for winter and the bees produced in the autumn have little work to do compared with their summer sisters and therefore have a longer life span, living throughout the winter until the annual cycle starts again.

3.1.1 Workers are produced from fertilised eggs (32 chromosomes and therefore female) laid by the queen in worker cells. When the egg hatches into a larva it is progressively fed with brood food, pollen and honey until the cell is sealed. It then turns into a pupa and finally into an adult bee when it emerges from its cell by chewing through the wax capping. The workers are female but not fully developed; they have no spermatheca for storing sperm and their ovaries are vestigial.

3.1.2 Queens are produced also from fertilised eggs (32 chromosomes) laid by the queen in queen cups or play cells. When the egg hatches into a larva it is fed continuously on a diet of royal jelly only until the queen cell is sealed. The feeding is always very generous in the case of the queen, shown by the remains of unused royal jelly which is generally evident in the bottom of the cell after the queen emerges. The development is similar to the worker with the larva turning into a pupa and then into an adult before emergence. The queen has a spermatheca and has the ability to lay, at will, either fertilised eggs for workers and queens or unfertilised eggs (16 chromosomes) for drones.

3.1.3 Drones are produced from unfertilised eggs (16 chromosomes and therefore male) laid in drone cells. The development is similar to the worker but with different timings.

3.2 The periods spent by the female castes and the drone in the four stages of their life (egg, larva, pupa and adult).

3.2.1 Definitions:

 •Metamorphosis - change in form by magic or by natural development or change (usually rapid)

between the immature form and the adult state.
•Caste (zoological definition) - form of social insect having a particular function.

It should be noted that the most recent syllabus for the examination only recognises two female castes and the drone. We believe the question of two or three castes is a matter of opinion, there being arguments in favour of both. As three have been recognised for over 100 years, the change seems unnecessary and we have not changed our original notes on the subject.

3.2.2 The three castes:

• Worker	-	derived from a fertilised egg (female)	32 chromosomes
• Queen	-	derived from a fertilised egg (female)	32 chromosomes
• Drone	-	derived from an unfertilised egg (male)	16 chromosomes

3.3.3 Stages in the life cycle:

	WORKER	QUEEN	DRONE
OPEN CELL:			
Egg	3d	3d	3d
Larva (4 moults)	5d	5d	7d
SEALED CELL:			
Larva/pro-pupa (1 moult)	3d	2d	4d
Pupa (1 moult)	10d	6d	10d
Total time from egg to emergence	21d	16d	24d
AS AN ADULT AFTER EMERGENCE:			
Summer bee	6w	3y	c.4m
Winter bee	c.6m	ditto	n/a

(c. = circa = approximately, d = day, w = weeks, m = months, y = years and n/a = not applicable).

Note that the above times can vary by a few hours before emergence due to variations in temperature of the brood nest.

3.2.4 Description of the stages in the life cycles.

Worker (before emergence):
　　1st day of egg- vertical, stuck to the bottom of the cell and parallel to the cell walls.
　　2nd day　　- at an angle of c. 45° to the cell walls.
　　3rd day　　- horizontal, egg lying on the bottom of cell. The egg hatches after 3 days.
　　4th-8th day　- the new larva starts eating immediately after hatching and starts to grow, moulting every 24 hours, until it fills the whole cell diameter. The cell is sealed on 8th day after the larva's last meal.

8th-21st day - the connection between the ventriculus and the hind gut opens and the Malpighian tubules open into hind gut; excreta enters hind gut and is voided into cell. The larva changes position and stretches out the full length of the cell (head outwards) and spins a cocoon. Metamorphosis occurs and the larva changes to a pupa after 5th moult 3 days after sealing. The pupa is still white but of adult form. It completes development, slowly changing colour and emerges from its cell by nibbling the capping on day 21. The 6th moult occurs just before emergence.

Queen and drone:

Similar but with the different timings as shown above.

Workers and drones are progressively fed with brood food for first 2-3 days, then with a mixture of brood food, pollen and honey. No food is consumed during the pupal stage. Queens are fed by mass provisioning with royal jelly throughout the larval stage. Queens are generally over fed and excess can usually be seen in the cell after emergence.

3.3 An elementary description of the function of the queen, worker and drone in the life of the colony.

3.3.1 Worker:

1st to 3rd day -	Cell cleaning and brood incubation.
4th to 6th day -	Feeding older larvae (brood food + honey + pollen).
7th to 12th day -	Feeding young larvae (brood food only).
13th to 18th day -	Processing nectar into honey, wax making, water evaporation and pollen packing.
19th to 21st day-	Guarding and starting to forage.
3rd to 6th week -	Foraging for nectar, pollen, water and propolis.

Note that these times are approximate and older bees can revert to their earlier duties if required by the colony. Other duties include ventilation, humidity and temperature control, etc. The duties undertaken by the worker bee are entirely dependent on its glandular development, for example a newly emerged bee is incapable of stinging because its sting gland has not developed. The first duties do not require any glandular functions and during the first three days the bee can only undertake menial tasks. During these first six days the hypopharyngeal glands (located in the head) develop and the bee is then capable of producing brood food ready for the next tasks which it undertakes. As the bee is itself consuming quite large amounts of pollen and honey so it simultaneously feeds older larvae. Consumption of honey stimulates the wax glands to develop fully and hence its next tasks are associated with wax making. The last glands to develop are those associated with the sting when guarding duties start. At this time the honeybee will be taking orientation flights to ensure knowledge of its locality prior to foraging duties in the field.

3.3.2 Drone:

Up to c. day 12 -	Generally confined to the hive except on fine days for cleansing and orientation flights.

33

| 12th to 14th day - | Mature and ready to mate (the drone's sole function). |
| Autumn - | Driven out of the hive to die. |

3.3.3 Queen:

1st day -	Seeking rivals and killing them.
3rd to 5th day -	Orientation flights to locate the hive.
1st to 3rd week -	Multiple mating flights.
Up to 3 to 5 years -	3 or 4 days after mating the queen starts to lay. Thereafter, she is solely egg laying and producing pheromones for colony cohesion until such times as the colony may swarm. The old fertile queen leaves with about 50% of the bees in the colony to form the swarm.

Note that up to the time the queen starts to lay, the times are not precise and are very much dependent on the weather. A queen that has not mated satisfactorily within c.20 days usually becomes a drone layer and is said to be stale. Such a queen is incapable of fertilising the eggs which she lays.

3.4 A simple description of wax production and comb building by the honeybee.

3.4.1 Wax production.

Wax is produced from glands on the worker honeybee; the queen and the drone do not have such glands. These glands are located inside the exoskeleton on the ventral or underside of the abdomen (on sternites A4-A7 inclusive). There are 2 internal glands on each ventral segment (sternite), making 4 pairs in all. The glands secrete a liquid which passes through the mirrors and oxidises as a flake of wax in the wax pockets. The glands, mirrors and pockets being known colloquially as the "waistcoat pockets".

a) Wax is secreted at relatively high temperatures (33°-36°C or 91°-97°F) after consumption of large amounts of honey. Various estimates are quoted for the amount of honey to metabolise 1lb of wax, 5-8 lb being a realistic estimate or 5 to 8 kg honey to metabolise 1 kg of wax.
b) Wax glands are best developed in worker bees 12-18 days old.
c) When building comb, bees hang in festoons near the building place, after gorging themselves with honey, waiting for the wax scales to form.
d) The wax glands inside the exoskeleton are covered with fat bodies and other cells.
e) The major components of beeswax are:

- hydrocarbons	16%	- fatty acids	31%
- monohydric alcohols	31%	- diols	3%
- hydroxy acids	13%	- other substances	6%

The chemistry of how the wax is produced and how it diffuses through the mirrors is extremely complex and it is not necessary to know the detail; however the candidate should be aware that a diffusion process is involved.

f)Wax is normally white but can be tinged with yellow hues caused by pigments that originate in pollen (eg. when a colony is working dandelion, new comb is noticeably coloured yellow).

g) For completeness a few of the physical properties of wax are:-

SG = 0.95.
Honey-like odour and a faint taste.
Pure wax melts at 146.9°F (63.8°C) and solidifies at 146.3°F (63.5°C).

3.4.2 Comb building.

The following points are relevant:

a) When building comb workers gorge themselves with honey and hang in festoons for c. 24 hours before the wax secretion and building process starts.

b) A wax scale is removed by one hind leg and transferred to the mandibles by the two fore legs. The wax scale is thoroughly masticated before fixing to the comb and moulding it in place. When it is first deposited it is spongy and flaky and is later manipulated again making it smoother and more compact.

c) Removing, masticating and fixing one scale takes about 4 minutes. On the basis of one scale taking 4 minutes to manipulate, about 66,000 bee hours are involved building 77,000 worker cells using 1kg of beeswax.

d) Bees can detect gravity (by using a hair-like sensor at the constriction or petiole between the head and the thorax) and the festooning chains of bees (catenaries) play an important role in the parallelism of the combs.

e) Queenlessness and bright light inhibit the bees from building comb and secreting wax.

f) According to Dadant, the thickness of the wall of newly built comb is approx. 0.0025in (0.064mm) thick and in naturally drawn comb without the use of foundation, the base is 0.0035in (0.0889mm) thick; Hooper gives 0.006in (0.1524mm) for the cell wall thickness and Winston 0.073mm (0.003in).

3.5 The importance of pollination to flowering plants and consequently to farmers and growers.

3.5.1 Definitions:

a) Pollination: is defined as the transfer of pollen from the anthers of a flower to the stigma of that flower or another flower on the same plant/tree or another plant/tree.

b) Fertilisation: is defined as the union of the male and female gametes which occurs after pollination.

3.5.2 The importance of pollination to flowering plants.

It is important to note that honeybees pollinate, they do not fertilise. When the flowers open and the stigma is receptive, nectar is usually secreted as an attractant to the pollinators. The aroma and colour of the flower is also attractive to the bee at a range of a few feet. There are a wide range of estimates on the number of pollen grains that can be collected on the plumose hairs of the honeybee; these range from 50,000 to 5,000,000 depending on which source is quoted.

There are a variety of methods of pollination which have been evolved by the flowers eg. insects, wind, animals, birds, water. The two following terms should be noted:

> anemophilous - wind pollinated
> entomophilous - pollinated by insects.

Honeybees evolved from wasp-like ancestors c. 20 million years ago, at the time the flowering plants were developing. A mutually beneficial relationship has developed between sexually-reproducing flowers and the honeybees. The bees provide for cross pollination of plants thereby ensuring a greater variety in the offspring than by self pollination. The plants in turn provide the bees with a reward of nectar and pollen. Other less important pollinators include flies, beetles, butterflies, bats and wind.

Honeybees have evolved branched hairs which can carry large numbers of pollen grains, intricate pollen baskets, specialised mouthparts, honey sac for transporting nectar, beeswax for building comb to store honey and pollen, specialised behaviour for communication, etc. all of which are related to their association with angiosperm plants (those that reproduce sexually).

The flowers attract bees by colour (flowers reflect UV light), scent, nectary guides and the shape of the flower.

The process of pollination is extremely simple; it is the transfer of the pollen grain from the anther to the stigma when the stigma is receptive and when the pollen is viable (alive). Without pollination there would be no fertilisation and no sexually-reproducing plants.

3.5.3 The importance to farmers and growers.

In UK most of the crops requiring pollination flower in the spring, the crop maturing in the summer and being ready for harvesting late summer/early autumn. The honeybee is the only pollinator which over-winters as a colony and which is available in the large numbers required early in the year. Practically all the solitary wasps and bees, together with the bumble bees and social wasps, start the season with only a fertilised queen who starts reproducing in the spring. This is the major reason that the honeybee is so valuable in the UK climate for pollination purposes.

Because the honeybee is kept in hives and managed by man, whole colonies in very large numbers can be transported and sited in the crops to be pollinated. The colonies can be distributed throughout the crop to the best advantage and provided in the optimum numbers required. The colony density for fruit pollination is about 2 - 6 colonies per acre.

Honeybees are polytrophic but constant to one plant while they are foraging (an advantage on most

36

crops for pollination purposes). A knowledge of bee behaviour is important for honey production and also for pollination.

It has been shown that crops such as rape (wind pollinated) provide a better set / greater yield if colonies of bees are available in correct numbers and correctly distributed while the crop is in flower. Any crop which relies on insect pollination produces better yields when bees are provided as pollinators. The UN organisation, FAO, has estimated that about one third of the human diet of the western world comes directly or indirectly from insect pollinated foods. About 90 crops farmed in the USA depend on bees for pollination.

3.6 The main local flora from which honeybees gather pollen and nectar.

The main flora will vary from district to district throughout the UK and for examination purposes the Examiner will only concentrate on the local flora. Below is a general list based on the flowering times; select the ones which are applicable to your area and your apiary.

COMMON NAME	PROPER NAME	FAMILY	NECTAR or POLLEN
February/March			
Snowdrop	*Galanthus nivalis*	*Amaryllidaceae*	P
Crocus	*Crocus spp.*	*Iridaceae*	P
Gorse	*Ulex europaeus*	*Leguminosae*	P
Hazel	*Corylus avellana*	*Corylaceae*	P
Willow (goat)	*Salix caprea*	*Salicaceae*	P
Yew	*Taxus baccata*	*Taxaceae*	P
March/April/May			
Blackthorn (sloe)	*Prunus spinosa*	*Rosaceae*	N+P
Dandelion	*Taraxacum spp* *	*Asteraceae*	N+P
Gooseberry	*Ribes uva-crispa*	*Grossulariaceae*	N
Currants	*Ribes spp*	*Grossulariaceae*	N
Rape	*Brassica napus*	*Cruciferae*	N+P
Top fruit	**	*Rosaceae*	N+P
Bluebell	*Endymion non-scriptus*	*Liliaceae*	N+P
Sycamore	*Acer pseudoplatanus*	*Aceraceae*	N+P
H. Chestnut (Wh.)	*Aesculus hippocastanum*	*Hippocastanaceae*	N+P
H. Chestnut (Red)	*Aesculus carnea*	*Hippocastanaceae*	N+P
Hawthorn	*Crataegus monogyna*	*Rosaceae*	N+P
Holly	*Ilex aquifolium*	*Aquifoliaceae*	N+P
Mountain ash	*Sorbus aucuparia*	*Rosaceae*	N+P
Laurel	*Prunus laurocerasus*	*Rosaceae*	N+P

June/July/August			
Poppy	*Papaver rhoeas*	*Papaveraceae*	P
Thistle	*Cirsium arvense*	*Asteraceae*	N+P
Hogweed	*Heracleum sphondylium*	*Umbelliferae*	N+P
Field bean	*Vicia faba*	*Leguminosae*	N+P
Raspberry	*Rubus idaeus*	*Rosaceae*	N+P
White clover	*Trifolium repens*	*Leguminosae*	N+P
Charlock	*Sinapis arvensis*	*Cruciferae*	N+P
Runner bean	*Phaseolus multiflorus*	*Leguminosae*	N+P
Lime	*Tilia vulgaris*	*Tiliaceae*	N+P
Blackberry	*Rubus fruticosus*	*Rosaceae*	N+P
Willow herb	*Epilobium angustifolium*	*Onagraceae*	N+P
Bell heather	*Erica cinerea*	*Ericaceae*	N+P
August/September			
Evening primrose	*Oenothera biennis*	*Onagraceae*	P
Ling	*Calluna vulgaris*	*Ericaceae*	N+P
Old man's beard	*Clematis vitalba*	*Ranunculaceae*	N+P
September/October			
Ivy	*Hedera helix*	*Araliaceae*	N+P
Michaelmas daisy	*Aster novi-belgii*	*Asteraceae*	P

* - *Taraxacum officinale* is the common dandelion; *spp.* denotes many species, in this case many species of dandelion.

** - Top fruit include apple, pear, cherry, plum, etc.; all are *Rosaceae*.

For the Basic examination it is not necessary to know the botanical names and family of the plants concerned. We believe that some students and candidates will wish to have a deeper understanding, therefore we have included the full scientific nomenclature (genus, species and family) in the table above. It should be noted that the classifications of some plants change from time to time (eg. Ragwort, *Senecio jacobaea*, family *Compositae*. The family is now re-named *Asteraceae*. In Scotland it has a common name 'Stinkywilly').

It is extremely difficult to define what are the main plants throughout the country because there are many local variations and the student or candidate should be familiar with his local flora; eg. winter aconite (*Eranthis hyemalis* of the buttercup family *Ranunculaceae*) is a prime source of pollen in the spring in some parts of Devon and for the last few years our own bees have worked Lesser Celandine (*Ranunculus ficaria*) at this time of the year. When the authors lived in Sussex, neither of these two plants were worked by our bees in the spring because they were not available.

It is to be noted that the labiates (eg. mint, thyme, rosemary, lavender, etc.) are all very attractive to bees but in UK they do not rate as major sources of forage (eg. thyme produces a crop of honey in Greece and Malta).

3.7 A simple definition of nectar and a simple description of how it is collected, brought back to the hive and is converted into honey.

3.7.1 Composition of nectar and its variations.

- Nectar is secreted by the nectaries of flowers and is composed of:

    a) water,
    b) sugars 5 - 60%, typically 20 - 40%,
    c) other substances - salts, acids, enzymes, proteins and aromatic substances.

- The sugars are principally sucrose(s), fructose (fr) and glucose (gl). The nectar types are discrete to each plant species which are generally in three categories, namely:

    a) sucrose dominant (eg. long tubed flowers such as clover, etc.),
    b) balanced nectar (roughly equal amounts of s, fr and gl),
    c) fr or gl dominant (eg. rape which is gl dominant like most crucifers).

- Bees tend to prefer balanced nectars but the nectar with the highest overall sugar content is usually collected in preference to one of lower sugar percentage.

- Gl/fr ratio of nectar:

    - high glucose content of nectar causes honey to granulate quickly with a fine grain.
    - high fructose content of nectar results in a honey which granulates slowly with a coarse grain.

    nb. glucose is less soluble in water cf. fructose or sucrose.

- Variations in nectar:

    a) composition by flower species - each is different,
    b) composition depends on:
        - weather conditions (temperature, humidity, wind speed, sunlight),
        - soil conditions (water content, PH, type such as chalk or clay, etc.).

- The variations in nectar secretion of plants is extremely complex with a large number of variables. Sunlight is very important as this is necessary for photosynthesis (hydrocarbons to nectar). With plenty of sunlight temperatures increase. A minimum temperature is required for the enzymes causing nectar secretion to operate. Rain may wash the nectar from the flower or dilute it, wind may dry the nectar evaporating some of the water and increase its sugar content; both conditions can sometimes be found on the same tree.

- Variations in the amount of nectar available in a given area is important to the beekeeper; only a finite quantity is available. An area can be over-stocked with bees to the detriment of the beekeepers concerned.

• A few figures on foraging areas and colony density:

a) Dr. Bailey considers that colony density should be no higher than 1 colony per 10 sq. kilometres to minimise disease (or 1 colony/4 sq. miles approx.).
b) The foraging area of 1 colony = πr^2 = 28.3 sq. miles (r=3 miles).
c) Therefore 7 colonies/apiary is about the maximum.
d) Compare this with 1 colony/ acre for pollination purposes or on a concentrated nectar crop such as rape. Note that 1 sq. mile = 640 acres.

It is important that the beekeeper should get to know his area; the flora, the micro-climates, other beekeepers and where each keeps his bees, all in relation to available nectar supplies.

3.7.2 The way nectar is collected and conveyed back to the hive.

The factors which encourage nectar collection have, surprisingly, been little studied. It is not known whether the amount of nectar collected is related to the amount of honey stored; it is obvious that collection extends far beyond the colony's actual requirements.

The presence of a queen and brood stimulates the collection of nectar in much the same way as it stimulates pollen collection. Unlike pollen, which is deposited directly by the forager into a cell, nectar foragers pass their load to a house bee and can also perform a wag-tail dance to recruit more foragers.

• Scouting and finding the source of forage occurs first. About 2% of the bees in a colony actually scout for forage. The scouts return with a load and communicate the source by dancing. The forage selected by the colony is likely to be the best in quality (highest sugar content) and quantity.

• Returning foragers are likely to repeat the wag-tail dances. The number of foragers in a balanced colony is about one third of the population (ie. two thirds of the adult bees are house bees).

•The bee sucks nectar up the food canal of the proboscis through the pharynx and oesophagus into the honey sac in the abdomen. The average load of nectar is 40mg (cf. the weight of bee = 90mg), the bee visits approx. 100 - 1000 flowers on each foraging trip. Foraging bees make c. 10 trips per day, each trip lasting 30 - 60 minutes.

•The enzyme invertase from the hypopharyngeal glands is added to the nectar as it transits the pharynx to the oesophagus. The conversion of sucrose (disaccharide) to fructose and glucose (monosaccharides) starts on the flight back to the hive. The process is continued by the house bee receiving the load from the returning forager bee.

3.7.3 Conversion of nectar to honey including chemical changes and storage of the honey by the bee.

The conversion of nectar to honey involves two changes:

a) chemical change (disaccharide to monosaccharides),
b) physical change (evaporation of water).

• Chemical change: A forager returning to the hive with a load of nectar transfers the load to a house bee which then undertakes the completion of the chemical change, which was started by the forager, as follows:

a) A small droplet of nectar is re-gurgitated into the fold of the partly extended proboscis and then swallowed (time about 10 seconds). This process is repeated 80 - 90 times in about 20 minutes on the same droplet which is then deposited in an empty cell or half-full cell. This re-gurgitation process = 'ripening'.
b) Sucrose is converted to glucose and fructose by the enzyme invertase from the hypopharyngeal glands added to the nectar by the house bee during the ripening process. Note that the nectar will contain all three types of sugar in varying quantities depending on the floral source (see section 3.7.1).
c) During the ripening process the water content of the nectar is reduced by approx. 15% as a result of evaporation when the droplet is exposed on the partly extended proboscis.
d) Finally the house bee undertaking the ripening hangs the unripe honey (now honey not nectar) to dry in either empty cells or half-filled cells.

• Physical change: is the process of evaporating the excess water in the unripe honey to bring the sugar concentration up to about 80%. This is done as follows:

a) A large amount of space is required (see section 4.6) as the honey is hung in empty or partially filled cells in order to provide the maximum surface area for evaporation purposes. In the empty cells, the honey is deposited with a 'painting action' on the upper surface of the cell.
b) Currents of air are distributed around the hive by the bees fanning at the entrance of the hive bringing in dry air and expelling moist warm air.
c) As the water content diminishes and the sugar concentration of the honey approaches 80%, the honey is moved and the partially-filled cells are completely filled and capped with a pure wax capping with a minute air gap beneath the capping.

3.7.4 Other points for consideration:

• It will be clear from the above that plenty of space is required and there is much sense in the old adage 'over super early in the season and under super late in the season'.
• It is important to provide conditions in the hive to allow the bees to ventilate and ripen their honey easily. The authors believe that by providing top ventilation it assists the bees to ventilate via the hole in the crown board and roof ventilators. In a nectar flow if the roof is raised there are always bees fanning around the open feed hole; we notice many beekeepers keep this hole closed for no apparent reason. It must be hard on the bees to move the air from the 3rd or 4th super down to the bottom entrance; no one would think of ventilating the bathroom by opening the front door!

3.8 A simple description of the collection and use of pollen, water and propolis in the honeybee colony.

3.8.1 The collection and use of pollen in the honeybee colony.

The colony needs a fertile queen and the pheromone from open brood to stimulate the foraging bees to collect pollen. Returning foragers recruit further foragers by dancing on the comb indicating to other bees the position of the source; the type of pollen is recognised by the aroma of the pollen on the bees' legs.

The number of pollen foragers can vary between wide limits depending on the colony requirements (eg. a few percent to as much as 90%).

When foraging, the bee alights on a flower and, moving quickly, bites the anthers of the stamen with her mandibles in order to dislodge the pollen grains. These pollen grains attach themselves to the plumose hairs which cover the whole of the exoskeleton. Then the bee leaves the flower and hovers nearby to clean the pollen from her body and to load it into her 'pollen baskets'. The process is as follows:

a) The front legs: by means of stiff hairs collect pollen from the head and first thoracic segment. The pollen is moistened by honey or nectar deposited on the front legs from the proboscis.
b) The middle legs: collect the pollen from the first legs and the rest of the thorax particularly the ventral side which is then passed on to the inner side of the basitarsi of the hind legs.
c) The hind legs: clean the abdomen and when sufficient pollen is collected on the inner surface of the basitarsi, these surfaces are raked by the 'pollen rake' at the bottom of the tibia of the other hind leg. The pollen is forced as a paste onto the flat surface of the auricle which is bevelled upwards and outwards. The tarsus closes against the tibia and the pollen is squeezed upwards and outwards onto the outside surface of the tibia. It is held in place here by the hairs on the corbicula, pollen basket, of the tibia (note the single hair acting as a pin through the load). One full load of two pellets represents approximately 100 flowers visited eg. dandelion when it is yielding well.

A few facts about pollen:

The average pollen load (both pellets) = 12 - 30mg.
The average trips per day = 6 - 8.
Total collected in one year = approximately 100 lb (45kg) per colony.
Amount required to raise one adult bee = 70 - 150mg.

• Storage of pollen. When the pollen forager returns to the hive with a load it has to be stored. She selects a cell near to the unsealed brood, grasps the edge of the cell with her forelegs and arches her abdomen so that the posterior end rests on the opposite side of the cell. The hind legs hang into the cell and the middle legs are used to push the pollen loads off the rear legs into the cell. The forager departs more or less immediately for another load. A house bee now comes along and breaks up the pollen and presses it firmly into the bottom of the cell with her mandibles. Honey or nectar is added to the pollen mass; it becomes darker, has a higher sugar content and is known as 'bee bread'.

42

The packed pollen can be fed to the brood or house bees (for producing brood food) or the cell can be filled with further loads topped off with honey and sealed with a wax capping for winter stores.

All pollen storage is adjacent to and around the brood nest where it is required for use though the odd cell or groups of cells filled with pollen are sometimes found in the supers.

It should be noted that after pollen has been collected by the bee, it is no longer viable for plant reproduction.

3.8.2 The use of pollen in the honeybee colony.

Pollen, the male germ cell of flowering plants (angiosperms), which comes from the anthers at the top of the stamens of flowers or catkins, has two major uses:

a) it is the principal source of protein, fat and minerals in the honeybee diet,
b) it can provide a surplus product from the apiary.

Pollen demand in the colony is related to the amount of unsealed brood. Bees cannot rear brood without pollen because the nurse bees cannot produce brood food from the hypopharyngeal glands unless the bees have consumed pollen. A strong colony will collect c. 50 - 100 lb (22 - 45kg) during a season.

a) It requires 70 - 150mg of pollen to rear one adult bee.
b) About 200,000 bees are reared during a season thus accounting for more than 50% of the income.
c) The balance is used by the adult bees preparing for winter (increasing their fat bodies) and/or stored in the comb for use early the following year before new supplies become available.
d) Note the weight of a worker bee = 90mg therefore 1pound of bees contains c. 5000 bees.

Pollen is rich in protein and is essential for body building material for growth/development and for the repair of worn out tissue. It also has the very important function of stimulating the development of the hypopharyngeal glands and the fat bodies of the winter bee. The protein content varies between different pollen types and also from flower to flower in the same foraging area. A protein content of c. 35% is typical of a high protein pollen eg. beans. Bees can discriminate between pollens by colour and odour; they cannot distinguish between the quality (protein content) of various pollens.

Pollen contains:		
	- proteins	7 - 35%
	- lipids (fats/oils)	1 - 14%
	- amino acids	
	- carbohydrates	
	- minerals	1 - 5%
	- vitamins	
	- enzymes	
	- water	7 - 15%
	- sugars	25 - 48%

43

There are wide variations in the content of different pollens and the bee more than likely receives a balanced diet due to the variety of pollens collected and used.

The use of pollen for brood rearing:

- Worker larvae are fed with brood food only from 0 to 3d. and then on 4th and 5th day with pollen, honey and brood food.
- Queens, both adult and larvae, are fed exclusively on royal jelly.
- After emergence of the worker bee, pollen is essential for it to reach maturity in a healthy state. It depends on pollen for the orderly development of its glandular system while it is a house bee.

In areas where natural pollen is in short supply, particularly in the spring, pollen patties can be fed to colonies to stimulate spring build up. Pollen shortage often occurs where colonies are foraging on honeydew in pine woods.

3.8.3 The collection and use of water in the honeybee colony.

Water is collected in the same way as nectar; it is sucked up with the proboscis and carried in the honey sac back to the hive. The collection of water in the honeybee colony is directly related to the use of honey and the collection of nectar by the honeybee colony. It is important to understand the mechanism which dictates whether water or nectar will be collected.

There are two liquid inputs to the hive, nectar and water.

a) water - used for cooling and humidity control of the brood nest and the dilution of honey.
b) nectar - used for immediate consumption by the bees and for converting to honey for storing.

Food transfer from bee to bee is a continuous process within the colony and the average honey sac content is 50% sugar and 50% water. This average content is important. It is the right mixture for digestion by the bee. The use of nectar, honey and water is shown in the 'Honey Usage' diagram in Appendix 4.

During wintering, water is required to dilute stored honey. Where does this come from? Winter water flights are very limited (cold weather). Other sources of water are:

- CO_2 + water vapour are expelled from trachea (breathing tubes of the honeybee). H_2O from this water vapour condenses within the hive,
- honey is uncapped by the bees and being hygroscopic it absorbs some water on the surface, diluting the surface honey,
- re-absorption through the walls of the small intestine of the honeybee.

Water collection is most noticeable in the spring when the bees are using large quantities of their honey stores. During the brief sunny periods early in the year, foragers are seeking pollen and water outside the hive as little nectar is available.

44

When water is used for cooling during the summer months, it is hung up in droplets and allowed to evaporate; this evaporation gives a drop in temperature, identical with the evaporation of water from our skin when we get out of the bath. Until we dry ourselves we feel decidedly chilly. Fanning is the first mode of cooling and ventilation used by the bees to maintain an even temperature in the brood nest and to expel CO_2.

3.8.4 The collection and use of propolis in the honeybee colony.

Propolis (pro = before, polis = city) is a resinous gum found on trees and other plants. It was given the name because of the instinct of many strains of bee to build curtains of propolis to restrict the entrance to their nests. Propolis is a resinous substance, orangey brown to red in colour, collected by honeybees from certain trees such as alder, poplar and horse chestnut together with the exudations from wounds in many woody plants (readily noticeable on cherry trees). It is made up of an indeterminate number of substances and therefore has no chemical formula.

It is collected with great difficulty by foraging bees in warm weather by:
    a) biting with the mandibles,
    b) kneading a small piece bitten off with the mandibles,
    c) transferring the piece with the 2nd leg to the pollen basket direct,
    d) patting it into position again with the 2nd leg,
    e) repeating the process to the other pollen basket on the other leg.

The load takes 30 - 60 minutes to collect. When the forager returns to the hive, it needs assistance to unload and the unloading takes place at or very near the site it is to be used. The shiny sticky loads of propolis are:

    a) unloaded by another worker bee,
    b) removed from the forager bee by biting and pulling the load which is then put in place where required,
    c) the 'cementing bee' may mix wax with the propolis,
    d) forager pats the remaining part of the load smooth again,
    e) forager is freed of its load in about 1 hour (or several depending on its use in the hive).

There are only a few propolis foragers in each colony (about 0.5% has been quoted) but this must vary considerably with the strain of bee and the availability of resin. Enormous quantities are sometimes collected eg. Caucasian bees.

The uses of propolis by the honeybee are:

    a) to fill cracks in the hive,
    b) to reduce openings eg. the entrance,
    c) to smooth the interior of the hive,
    d) to varnish the interior of brood cells; in this respect it should be noted that propolis has anti-bacterial properties which prevent the growth of bacteria in the brood nest,
    e) to strengthen comb attachments,
    f) to cover intruders, when they are dead and too large to carry out of the hive eg. mouse, slug etc.

The uses of propolis by man are:

    a) medicinal and veterinary work,
    b) as a varnish or shellac eg. on violins.

The beekeeper dislikes propolis because:

    a) it sticks to the hands, clothing etc. requiring methylated spirit to remove staining,
    b) it contaminates beeswax,
    c) it is difficult to remove from sections for sale,
    d) it makes frames and other moveable parts of the hive difficult to remove. It is however useful for migratory work by keeping the hive parts stuck together.

3.9 An elementary description of swarming in a honeybee colony.

The syllabus fails to mention supersedure and an account of swarming in a honeybee colony would be incomplete without reference to this very important aspect of honeybee behaviour.

• Swarming and supersedure are survival and reproduction instincts. It is not certain whether swarming developed from what we know now as a mating swarm or whether it evolved from absconding in times of dearth.

• Supersedure is the changing of the queen in a colony without swarming taking place; swarming makes provision for a new queen in the colony if one does not exist before the swarm departs. We believe that in general about 10 - 15% of all colonies supersede every year.

• The reasons for swarming and supersedure:

    1. An inadequate supply of 'queen substance' to all the workers in the colony is the fundamental reason why this happens.
    2. With an ageing, damaged or sub-standard queen which can still be laying.
    3. Due to congestion (overcrowding in the available space) which causes a breakdown in the food transfer mechanism and deprives some of the workers of queen substance although the queen may well be producing adequate quantities.

When 'queen substance' is in short supply ie. when workers are not receiving the minimum threshold amount, queen cells are produced by the workers and when this happens 3 things can occur:

    a) swarming can take place,
    b) the old queen can be superseded,
    c) the bees will do neither and tear down (destroy) the queen cells they have built.

• It is unknown how the colony decides to proceed. When queen cells are started, the beekeeper must assume that the colony will swarm, particularly in the season before the main flow.

• After eggs have been laid in queen cups, the queen is given less food and egg laying falls off. The queen's abdomen contracts and she gets lighter in weight in preparation for the swarming flight. The queen would have difficulty flying when she is in full lay.

• After the sealing of the first queen cell, the first swarm (prime swarm) will issue from the colony usually between 10.00 - 15.00 hours, depending on the weather. This prime swarm contains the old queen plus about 50% of the colony's bees. If the weather has been continually bad for 8/9 days after sealing of the first queen cell, then one or more virgins may also be included in the swarm.

• A second swarm may issue when the first virgin emerges, usually 8 days after the emergence of first swarm. This may be followed by a further cast unless the first virgin out destroys the other queens either in their cells or by fighting the newly emerged virgins. These 'casts' are very much smaller than the prime swarm, sometimes only c. 1500 bees plus one or more virgins. The origin of these casts may be mating swarms whereby the bees leave with the virgin queen. At the issue of each cast about 50% of the remaining bees depart; it will be clear that if a prime swarm plus one cast issues from one colony, its size is seriously eroded and chance of survival endangered.

• Prime swarms usually emerge and settle close, a few metres, to the old hive and stay for 12 - 48 hours before taking off for their new home. Casts on the other hand are capricious and seldom cluster for long in any one place.

• Swarming generally occurs in May and June (S. England) but early flora can cause a rapid build up and swarms can occur in April (nb. rape).

• Supersedure generally occurs in August and early September. Often the old and the new queen can be found in the same hive (for a short period), sometimes both laying, until the new queen takes over and the old one is rejected. It is not clear whether the new queen kills the old one or whether the bees ball and eject the old one from the hive.

• If the original cause of swarming was insufficient queen substance, the new colony may well be a stable unit ie. fewer bees for the same amount of queen substance produced by the queen giving each bee the required threshold amount. There is however a strong possibility that this new colony, formed from the prime swarm with the old queen, will supersede later in the season.

• When a swarm settles it has 2 distinct parts:

   a) an outside shell bees about 2 to 3 inches thick (providing protection and mechanical strength),
   b) an inner part (loose) consisting of chains of bees connected to the shell.

The outer shell has a distinct entrance to the inside. Dancing (wag-tail) can often be observed on the outside shell indicating the selection of a new home for the swarm.

• The bees in a swarm, before emerging, will gorge themselves with honey ready for wax building

47

at short notice when the new home has been found. For this reason swarms are generally very docile, however care should be exercised with swarms of unknown history. The stores consumed before the swarm departs are the only resources available, carried from the old colony, to establish a new nest at a new site.

3.10 An elementary description of the way in which the honeybee colony passes the winter.

3.10.1 The origin of thermoregulation in the honeybee.

The origin of the honeybee, *Apis mellifera,* was in Central Africa. During its evolutionary period of about 30 million years it successfully migrated to the northern latitudes. In so doing, it developed a way of flying in cold weather, keeping sufficiently warm in winter to survive and collecting enough stores until they became available again the following year. The origin of its ability to keep warm has developed from being able to fly at low temperatures.

- The heat losses from the exoskeleton are generally low and the honeybee is poikilothermic (taking the temperature of its surroundings, ie. the ambient temperature).
- Its thoracic temperature (T_{Th}) is about 10°C (60°F) above the ambient temperature (T_{Amb}) when not flying.
- In order to fly, T_{Th} must be greater than 27°C (80°F) because at lower temperatures the minimum critical wing beat frequency cannot be generated ie. the minimum frequency for take off (to get lift).
- This puts a constraint on the enzymes working in the flight muscles which will work at high T but not at low T. This is a characteristic of most enzymes, eg. fairly high temperatures are required for the enzymes in the nectaries of a flower to work in order to secrete nectar.
- The honeybee has developed a method of warming up these flight muscles prior to take off by operating simultaneously the wing elevator and depressor muscles in opposition to one another creating no movement but using energy and generating heat until the required critical temperature is reached ($T_{Th} = 27$°C).

It is this ability to generate heat with no apparent movement that provides the means for nest warming in cold climates not only in the winter but in the temperate summer temperatures of northern latitudes.

Honey flows in the tropics, the origin of the honeybee, are two per year and generally coincide with the monsoons. In the subtropics (eg. the Mediterranean) the honey flow is more or less continuous throughout the year which is why the yellow bee with its roots in southern Italy continues to rear brood into early winter and uses up its stores in the process if kept in northern latitudes. In northern latitudes there is generally only one main flow. The honeybee which has evolved for these climatic conditions builds up on the spring flow and then stores a lot of honey in a short time on the main flow and reduces its brood production when the flow comes to an end, thereby conserving its stores for winter use. It is to be noted that the production of oilseed rape is tending to distort this pattern, to some extent affecting the seasonal colony development.

The last feature that enables the honeybee to survive the cold winters in high latitudes is its ability to vary its age and instead of having a life span of 6 weeks in the summer this is extended to 6 months in the winter ie. the summer and winter bees. In common with many other animals the honeybee develops fat bodies in large quantities in preparation for winter. Additionally, the hypopharyngeal glands are highly developed in c. 80% of the winter bees as compared with c. 20% of the summer bees. There is a marked variation in the content of the fat bodies as follows:

WINTER BEE	high protein, low fat
SUMMER BEE	low protein, high fat

A further feature of the ability to generate heat is for the incubation of brood. The brood nest temperature has to be maintained between very close limits with a mean of 90°F (33°C). If the brood nest temperature drops, even a few degrees, to 85°F or 86°F deformed bees will result with stunted abdomens and deformed wings. Little research work has been done on the aspects of low temperatures in relation to disease but there are indications that low temperatures could be the trigger for the development of disease in the colony eg. Chalk brood, *Ascosphaera apis*, spores need low temperatures to germinate.

3.10.2 How the colony prepares for winter.

a) It collects a summer/autumn harvest.
b) Drones are evicted usually after the main flow in August/September.
c) Dramatic loss in the summer bees.
d) Large increase in the winter bee population resulting in:

- Consumption of large quantities of pollen and honey.
- An increase in size of the hypopharyngeal glands.
- A large increase in the number of fat bodies which store glycogen (carbohydrate from honey), protein and fat.
- Longer life as a result of no work to do rearing brood (lower CO_2) or foraging.
- Life of bee $\propto$ pollen consumed $\div$ (brood reared $\times$ CO_2 level).

3.10.3 How the colony over-winters.

Winter is characterised by the absence of flowers and the consequential lack of pollen and nectar for the bee to forage. The days decrease in length and the ambient temperature decreases. The reaction of the honeybee is as follows:

- The queen stops laying not abruptly but slowly reducing to nil c. October/November and starts again very slowly at first c. January/February.
- The honeybee is poikilothermic, it is immobilised at 45°F (7.2°C) and starts to die at 40°F (4.4°C). The only way it can survive is by clustering. The characteristics of the cluster are as follows:

a) The cluster starts to form at 57°F (14°C) always below the food stored.

b) The outer shell is 1 inch to 3 inches (25mm to 75mm) thick with a partially filled centre of festooning bees.

c) The outermost bees forming the shell enter the empty cells in the comb.

d) From 57°F to 45°F (14°C to 70°C) there is little change in the size of the cluster, the heat losses are regulated by heat produced by the bees.

e) 45°F (7.20°C) is the critical temperature (T_{cr}) and if the temperature of the bees on the outside of the cluster dropped lower they would become immobilised and drop off. Below T_{cr} the cluster contracts and above T_{cr} the cluster expands.

f) As the temperature continues to fall the bees on the outside of the cluster bury their heads into the cluster and spread their wings forming an efficient heat protecting layer. Heat is generated inside the cluster to maintain the surface temperature at no lower than 45°F (7.2°C).

g) Considering the heat losses from the cluster, clearly heat lost = heat generated by the bees. The heat lost can be by conduction, convection and radiation . Conduction loss is virtually = zero; wooden frames and wax comb are poor conductors of heat. Convection losses due to air currents around the cluster are approximately = to the radiation (mainly infra red) losses. It follows then that the smaller the cluster (minimum surface area) the smaller will be the convection and radiation losses.

h) Consider the surface area of a bee = $2cm^2$. If there are 15,000 bees their total surface area = $30,000cm^2$. If they now form a cluster 18cm radius the sphere has a total surface area of $1000cms^2$. This represents a saving of × 30 in surface area and a corresponding saving in heat loss. If the cluster contracted further to 12cm radius the saving factor = × 66; a very efficient method of conserving heat and minimising energy used.

i) It is essential that the cluster maintains contact with the food reserves throughout a prolonged cold spell. It sometimes happens, but not very often, that a colony does get separated and starves after it has drawn on the individual fat body reserves of each bee.

j) It is to be noted that there is usually a connective cluster of a few hundred bees making contact with the food reserves and said to direct the main cluster around the combs during the clustering process. We have not been able to locate much meaningful information about this aspect of wintering behaviour.

k) As the days start to lengthen after the winter solstice (December 21st) the queen starts to lay again. We believe that the lengthening days are the trigger to start feeding the queen again and bring her back into lay. This behavioural aspect has not been proven and there is a paucity of information about it in the classical literature.

l) By clustering and generating heat a normal healthy colony of *Apis mellifera* from northern latitudes with adequate stores can survive temperatures down to -31°F (-35°C) without any special management by man.

** ** ** **

4.0 BEEKEEPING

4.1 An elementary description of how to set up an apiary.

4.1.1 General considerations.

Before the siting of colonies in an apiary, consideration should be given first to the site itself and indeed whether it is suitable for keeping bees. Criteria for the site are as follows:

a) Is there adequate forage in the surrounding area for the colony to support itself and can it readily obtain water throughout the year?

b) There must be no question of danger to humans, particularly children, or animals. One sting can kill an allergic subject if not given expert medical treatment very quickly.

c) Ideally, the apiary should be in a place where nobody except the beekeeper can be stung. This criterion is nigh impossible to achieve and siting of the colonies in the apiary becomes of vital importance to minimise this risk.

d) Under no circumstances should an apiary be established adjacent to a public thoroughfare, even if there is a barrier (eg. hedge or wall) of suitable height between. Bad tempered bees while being manipulated will attack moving human and animal targets for quite large distances from their hive.

e) The site should not be in a frost pocket and protection from the prevailing winds is most desirable.

f) The site should be free from any form of flooding, under trees in or on the edge of a wood.

g) The site should be accessible by road at all times of the year.

h) The site should be surrounded by a stock-proof fence if it is adjacent to pasture where livestock is likely to be grazing.

i) Finally a matter of ethics relating to out apiary sites which are often on farmer's land; if another beekeeper has bees close by on the same owner's land, then look elsewhere for another site no matter what the farmer says.

It cannot be over emphasised that the utmost care must be taken in siting an apiary and the colonies in it. Bees cannot be moved around like other livestock. If there is any doubt about any aspect of siting, expert advice should be sought.

4.1.2 Detailed considerations for siting colonies in a small home apiary.

A small home apiary implies that there is a dwelling house nearby with other people, children and domestic animals, if not in the in the same grounds then on neighbouring property. The same or similar considerations apply to an out apiary, therefore the following points require attention:

a) Stocks must be sited so that the flight path of the bees avoid footpaths and areas where there is likely to be any human or animal activity. Stocks can be sited so that bees have to fly up and over hedges and fences thereby getting the bees to a safe height above anyone on the ground. Under normal circumstances such an arrangement is quite workable but aggressive bees must be considered. It is absolutely essential to have a 'bolt hole' prepared over 3 miles away, so that a stock may be moved in an emergency.

b) There must be plenty of space around each stock for colony manipulations and maintaining the site (eg. grass cutting). A distance of 6ft. between colonies would not be out of the way for setting up nucs and doing artificial swarms between the adjacent stocks.

c) Space should be allowed at the planning stage for expansion in the future; this aspect is often overlooked.

d) The layout of the stocks should be in an irregular fashion in order to minimise drifting.

e) Hives should be provided with permanent bases to raise the floorboard off the ground to prevent damp and possible rot starting to occur in the lower woodwork of the hive. Concrete bases are undoubtedly the best but try a temporary solution until the site has been tried out for a couple of years.

f) The height of the top of the brood chamber, to minimise too much bending, is very important if large numbers of stocks are involved. Even six hives become a real pleasure if they are at the right height compared with being too low. A point to consider when designing hive stands.

g) In home apiaries it is best to site the stocks out of sight of neighbours if this can be done.

h) Bees in the stocks will at some time swarm despite the best efforts of the beekeeper to prevent this happening. Shrubs and trees around the stocks are useful for the swarms to hang on.

i) A certain amount of shade from nearby trees is useful particularly at mid-day during the summer. Stocks should not be sited under trees where rain drops can fall from them onto the hive in winter; it disturbs the colony.

Provision should be made for storing spare equipment near to the apiary, preferably a discrete shed for the multitude of bits and pieces. There is a risk to life, albeit fairly remote, wherever bees are kept and if this is remembered when planning an apiary, the chances of success are well assured. Having said all this, we believe that suburban gardens are becoming so small and houses so close together that they are unsuitable sites for keeping bees and certainly not the place for beginners.

4.2 Precautions that should be taken to avoid honeybees being a nuisance to neighbours and livestock.

Nuisance is caused in a variety of ways but the three most important are stinging, swarming, fouling newly washed clothes put out to dry and soiling parked cars. When neighbours are stung it is generally as a result of the bees being of doubtful temperament

4.2.1 How the variable temperament of bees arises.

In order to address this part of the syllabus it is necessary to have a clear understanding of why the temperaments of our honeybees are variable and how this arose. There is a very wide range of temperaments from the very docile to the very aggressive or, put another way, from those that exhibit little defensive instinct to those with a very strong defensive instinct, whether the colony is disturbed or remains undisturbed.

By various scientific techniques, including palaeontology, it is generally agreed that the origin of the honeybee was in that part of Africa that is now known as Kenya somewhere between 20 and 30 million years ago. This was before India split off from Africa and before the Red Sea and Rift Valley were formed. The land mass now called India carried with it some of the earlier bee-like insects which developed into the Eastern species of bees, ie, *Apis dorsata*, *A.cerana* and *A.florea*.

Reference now to the diagram of the 'Migration of the Honeybee' in Appendix 1 shows the origin in Kenya and the main migration routes marked with double line arrows northwards, southwards and westwards. Major races developed notably *A.m.capensis* in the Cape of Good Hope area, *A.m. adansonii* in the equatorial strip, *A.m fasciata* (the Egyptian bee) in the north east and *A.m.intermissa* (the Tellian or Arab bee) in the north west.

The Tellian bee is considered to be a major race from which many other strains have developed. In its native part of northwest Africa it developed in a very hostile environment and presumably adapted to these conditions by having to defend its colonies against determined predators eg. hornets, etc. It migrated northwards before the last melting of the ice cap c. 10,000 years ago and established the well known races in NW Europe such as the Iberian bee, the French Black bee and the English bee. It migrated as far north as Finland in latitude 60^0 N.

Similarly, the other major race (*Fasciata*) migrated northwards also providing the Italian bees (*ligustica*), Greek, Caucasian Carniolan, etc. After the ice cap melted the UK was cut off from the continent and the Mediterranean was formed, isolating Europe from Africa. At that time we had a discrete race of bee in the UK now known as the English Black bee or the English Brown bee. It was decimated by disease in the early 1900s (nb. Isle of Wight Disease or Acarine) and the Government of the day offered subsidies to beekeepers to import bees from the Continent. The result was that most of the races in Europe were imported and over the last 50 to 80 years these races have interbred and we now have a hotchpotch of mongrels with a very mixed pool of genes.

All the bees with their origins emanating from the Tellian bee have to a greater or lesser degree the defensive instinct, while other bees from Italy, Greece, etc. have a very weak defensive trait. Breeding can only be true with pure strains, while breeding with mongrels is known to be an erratic and unpredictable procedure.

Herein then lies the root of our problem in the UK. Using mongrel bees and taking pot luck with the matings will result in a wide range of temperaments in the offspring. The only beekeeper who claims experience with the old English bee, and has committed his findings to paper, is Bro.Adam. He considered the bee had some very desirable features but also some very bad ones, notably bad temper. BIBBA claim that there are pockets of the old English bee with a good temper and campaign for its breeding and use in the UK.

Before leaving this introduction to understanding the temperament of our bees it must be pointed out that other characteristics will also be present in mongrels, such as high tendency to swarm, heavy propolis gatherers, early starters/late finishers, longevity, economy on stores for overwintering, resistance to different diseases, etc. We know from experience that it is possible to eliminate bad temper fairly quickly by culling those queens producing bad tempered bees. However, it is nigh impossible to control more than two variables without recourse to specialised isolation and mating techniques. Our own breeding programme concentrates on good temper and minimum swarming tendency which can be achieved by any beekeepers who put their minds to it. Unfortunately, it is a very small minority who control the temperament of the bees they keep and we put this down to two traits that have developed over the last 50 years as follows:

a) Newcomers have been encouraged to wear very adequate protective clothing which has become available, making them 'safe' from stings under most conditions.

b) Few associations encourage their membership to manipulate their bees without gloves.

4.2.2 The defensive qualities of different strains.

We may well ask how we can measure the defensive qualities of the bees we keep and the bees we wish to get rid of?

Lord Kelvin (1824 to 1907) stated on one occasion "I often say that when you can measure what you are talking about and express it in numbers you know something about it; but when you cannot measure it in numbers, your knowledge is of a meagre and unsatisfactory kind".

We are all aware that it is possible to open up some colonies of bees without a veil or other protection and it is unlikely that a sting will be suffered by the operator, assuming that he knows what he is doing and handles his bees well. We are also aware that other colonies are virtually impossible to handle. In between these two extremes is the wide variation of temperament. The best measurement to assess the defensive nature of the bees is, in our opinion, to take note of the followers. If, after manipulating a colony, there are no followers 2 metres from the back or side of the hive then it is a suitable bee for keeping in a suburban garden. At 2 metres one should be able to remove the veil and not get pestered by a guard bee or follower.

If bees follow up to 5 metres then, in our opinion, they are unsuitable for a suburban garden. Such bees will need careful handling during a manipulation and would only be suitable in an isolated out apiary.

If bees follow beyond 5 metres then we would consider requeening them with a more favourable strain.

We consider that the use of gloves should be dispensed with except for dire emergencies. If the bees cannot be handled without gloves, then there is something wrong with either the bees or the beekeeper, both of which can be corrected. We record the number of stings we receive from each colony on the record card and this is also a numerate way of measuring the defensive qualities of the stocks. We don't expect to get stung during an inspection but if we do then there must be a reason for it and we do our best to assess the situation. If we wore gloves as the norm we would be unlikely to assess in this way.

Finally on this point, if the colony is becoming difficult to handle then close it down; contrary to what many experienced beekeepers do, it is very good beekeeping practice.

4.2.3 Variable temperament in relation to management.

The more bad tempered the bees the longer it takes to undertake routine inspections for swarm prevention and control. Additionally, the enjoyment of managing a few colonies of bees decreases rapidly with even a modest amount of bad temper.

When the colony gets to a state where it is really difficult to handle, the average beekeeper gives up and regular inspections go by the board. Such a beekeeper then starts extolling the virtues of leave alone beekeeping which is usually nonsense. Swarms start to issue and the result is that the bad temper is further promulgated around the area.

Even the best beekeepers occasionally end up with a rogue colony and if this is in a garden apiary something has to be done at very short notice. All beekeepers must have a 'bolt hole', a site where a bad tempered colony can be moved safely away from children, neighbours and livestock where it can be dealt with at leisure and put to rights. Here it must be appreciated that the bad temper may be due to the genetic make-up of the queen's offspring or due to the pheromones produced by the queen. If the cause is genetic then it will take a minimum of six weeks after requeening for the temperament to improve; three weeks for the house bee and then another three weeks foraging before it naturally dies. This is a long time during a relatively short season. If, on the other hand, the cause is due to pheromones (our guess at the cause and not proven) the change will be effected in 24 to 48 hours. To the best of our knowledge there has been no scientific work done on this aspect of pheromones and temper. However, by experience we know it to be true and it can be readily demonstrated by reversing the queens. We have experimented with nuclei by swapping queens.

If colonies are left to produce their own queens and mate in the local neighbourhood it is surprising how quickly they deteriorate and up surges the bad temper. In 1986 we moved to Devon and brought 6 well- behaved colonies with us for the garden apiary. Other more pressing demands on our time precluded the continuation of our selective queen rearing programmes. In three years we had six very 'grotty' stocks which had to be moved to an out apiary which we had found and established by that time. It is essential for the management of good temper that queens are reared each year and the best selected for use in honey-gathering colonies.

The only way that we know, using mongrels, is to rear from the best tempered stock in May and overwinter the queens in nucs for introduction the following spring. By the time spring has arrived the queens will have been assessed for first temper and then nervousness. If they are satisfactory they can be used. The old queens, no matter how bad they are, can be used to keep the nucs going until queen rearing time again. Failing this, it is necessary to buy in queens of known temperament.

We believe that only about 5% of beekeepers do any selective queen rearing with the result that bad temper is a perennial problem.

4.2.4 Public relations and the temperament of bees.

Public relations and the temperament of bees is usually confined to bees and neighbours in the suburban situation. The bees kept adjacent to neighbours must be beyond reproach as far as temper is concerned and they must not be of a swarming tendency. No matter how nice ones neighbours, our bees must not cause nuisance by stinging, swarming or soiling the washing; all are unacceptable.

In all the years we have kept bees in our garden, only once has a neighbour been stung. She was about 6 feet from one of our hives weeding her garden the other side of an open mesh fence. She said it didn't matter but we told her it did matter to us and the offending stock was requeened although it was not, in our opinion, a bad stock.

It is important to manipulate your bees in a garden without gloves if you are overlooked by neighbours, it tells them that the bees are unlikely to be a menace to them. It is even better if the bees can be kept out of sight from one's neighbours but always ensure that you approach the hives without a veil and similarly when leaving the bees.

It is good practice never to open up or manipulate a colony in a garden apiary when the next door neighbours are using their garden.

Good public relations and bees are just plain common sense but do ensure that insurance is kept valid for third party liability. Litigation seems to be the order of the day and attitudes can change dramatically if a serious incident occurs.

4.2.5 The actions which can be taken to avoid bad-tempered bees causing a nuisance to members of the public.

The actions are common sense and are as follows:

1. Select the apiary carefully in accordance with section 4.1.
2. Site the colonies in the apiary carefully, also in accordance with section 4.1, ensuring that flight paths do not interfere with any member of the public.
3. Ensure that colonies of good tempered bees are maintained as detailed above.
4. Move any offending bad tempered bees immediately or, if necessary, kill the stock off.
5. Requeen if the situation is not immediately serious.

4.2.6 Requeening bad-tempered colonies.

Always the problem is to find the queen and it will be a queen that is unmarked. In one's initial years of beekeeping seek out the assistance of an experienced member of your association to help you. This is what the branch associations are all about. Remove the colony if at all possible to your pre-planned 'bolt hole' is the best advice.

It is always best to have a look at the colony quietly to see if it is really bad tempered. We have had many calls only to find that we could handle the colony without gloves and most of the trouble is the way the beekeeper manipulates the bees. Bad smoker technique, clumsy handling, etc.

By moving the colony. Early in the morning move the colony away from its normal position (about 8 feet) and put an empty brood box with about six combs in it to catch the foraging bees on return. Remember about 33% of the colony are foragers and sufficient comb should be provided. At mid-day when all the foragers have gone find the queen and introduce the new one. Next morning swap the boxes over, returning the hive to its original stand and the foragers in the place where the hive was temporarily placed. Remove the empty box of frames in the evening of the second day. Four days later release the queen, if the release is to be supervised, or remove the cage if released by the bees.

56

By using chloroform. This is only for the really bad bees! The colony that sends the bees downwards into your 'wellies' and finds every crack in your armour. Kit up well and carefully because the returning foragers are the problem but these can be avoided by shifting the stock as above.

The method is due to the late Harrison Ashforth, CBI for Cornwall. It works well and we have used it on quite a few occasions on other peoples bees. About 1 fluid ounce of chloroform is required, two pieces of corrugated cardboard about 3 inches (75mm) square and a piece of foam to seal the entrance. The method is as follows:

- The entrance block should be in the colony. Drive the bees back from the entrance with smoke.
- Pour one teaspoonful of chloroform on to one of the pieces of cardboard, push it into the entrance and seal with foam or a rag.
- Do exactly the same with the second piece of cardboard putting it under the crown board; the feed hole should be covered to keep the fumes inside.
- Wait for 2 to 3 minutes.
- You then have about 10 minutes maximum to work and find the queen.
- Remove her and run another one in directly from a cage laying on the top bars.
- As soon as the new queen has walked in close up and open the entrance. The job is complete.

The chloroform affects the central nervous system of the bees and disorientates them. We assume the mechanism is that in a disorientated state they cannot recognise their own queen and immediately accept the new one. The method has, so far, never failed and Harrison Ashforth also reported similar results with no failures.

4.3 The possible effects of honeybee stings and able to recommend suitable first aid treatment.

4.3.1 General.

- It is the mature worker bee, 14+ days old, which is capable of injecting bee venom into its chosen victim. Queens only use their sting on other queens and drones have no stinging apparatus. It is a means of defence used by the honeybee, generally as a last resort. After stinging mammals, the honeybee leaves behind the stinging apparatus, including the 7th abdominal ganglion, thus terminating its own life.

- Stings can be minimised as follows:

a) Beekeepers maintaining stocks of docile bees by culling the queens producing nervous bees and those with strong defensive traits eg. followers.
b) Handling colonies correctly i.e. no jarring of frames, no fast movements, no squashing of bees or banging of hives. Bees are very sensitive to vibration.
c) Only open stocks under good weather conditions. Thunderstorms or approaching rain clouds definitely affect the temper of the bee.
d) Always wear a veil to protect the eyes, the nose, the mouth and ears where there is a

proliferation of mast cells (mast cells are pharmacological 'time bombs'; when they rupture they release powerful chemicals that can effect various tissues nearby, such as blood vessels or smooth muscle).

e) Use protective clothing of correct material eg. cotton. It is said that woollen garments or garments dyed blue are best avoided. In our opinion the temperament of the bee is the predominant factor and not the clothing.

f) Dispersing the sting pheromone by the application of the smoker to the site of injection does discourage other guard bees being attracted to the site of the first sting. For stings on the hands, place a hot part of the smoker on the sting area to evaporate as quickly as possible the volatile pheromone and then smoke the area to mask any remaining smell.

g) Refrain from using perfumes, aftershave lotions, hair shampoos, nail varnish, hair sprays and other similar cosmetics prior to working with bees as these have been known to elicit a stinging reaction from bees due to the similarity in the chemical make up to the isopentylacetate in the alarm pheromone of the sting chamber.

h) Site stocks as far away as possible from the general public.

4.3.2 The effect of stings.

• Most beekeepers at the first sting may experience pain, reddening of the skin and swelling. The severity seems to vary depending on the site and the number of stings. But over the years a natural immunity will be built up and the discomfort and swelling will be minimal.

• Extensive swelling or irritating rash may occur taking 12 hours to reach its maximum and 2 or 3 days to resolve. These symptoms may indicate an increasing sensitivity to bee venom.

• A generalised reaction with symptoms of difficulty in breathing, skin rash, palpitations, vomiting and faintness occurring within minutes of a sting indicates a severe reaction (anaphylaxis) requiring emergency medical attention.

4.3.3 First aid treatment.

• The barb should be removed as soon as possible by a scraping action of a knife, hive tool or finger nail, not by squeezing between the fingers. This is likely to inject more venom by squeezing the bulb which contains the reserve of venom.

• Away from the stocks of bees, applying ice to the site of the sting, especially if it is situated where there is little spare skin for expansion eg. ear, nose or tip of finger, will bring some relief of pain.

• Application of calamine lotion, steroid creams or crushed leaves of the marsh mallow plant (*althaea officinalis* has long been used to relieve inflammation) may give some relief. Antihistamine creams are best avoided as their repeated application can cause severe skin sensitization.

• Aspirin tablets may reduce pain and inflammation. Piriton tablets contain antihistamine and may lessen the symptoms. If in doubt consult a doctor.

• To the non-beekeeper a sympathetic attitude will often solve the immediate problem but should there be any severe reaction, as may be the result of a sting close to or on the eye ball or in or around the mouth or neck, then it is always safest to obtain medical advice.

4.3.4 Other points of interest.

• Most beekeepers expect to receive the odd sting and take no further action beyond removing the barb.

• About 15% of the population have an atopic constitution and roughly 50% of the children of two atopic parents will be similarly afflicted. This group includes individuals with a personal history of hay fever, eczema, asthma, allergic rhinitis and urticaria. They may show progressive worsening in their reactions to stings developing general symptoms such as nausea, skin rashes and respiratory difficulties. Medical aid should be sought immediately should anyone show these symptoms.

• Taking non inflammatory drugs eg. aspirin or piriton under medical advice one or two hours before working in the apiary does reduce the reaction to stings. It should be remembered that antihistamines cause drowsiness so that driving the car is unwise after this kind of medication.

• For the hypersensitive person who wishes to continue beekeeping or for the members of the beekeeper's family who are exposed to the danger of being stung and are hypersensitive a course of immunotherapy can be arranged through their family doctor.

4.4 An elementary description of the year's work in the apiary and of the management of a colony throughout a season.

The beekeeping year is generally acknowledged as starting after the main crop is removed the previous year. The year's work in the apiary is examined below on a month by month basis starting in August on the assumption that the main flow occurs in July and all the supers have been removed by the beginning of August. Included in the account below are bits of information outside the BBKA syllabus but which all good beekeepers should know.

4.4.1 August: This is the month that work commences to prepare the colonies for winter.

4.4.1.1 General. Preparations for winter should start in August after the main crop has been removed and extracted. There are reasons for this:

a) Under normal circumstances a colony of bees collects all the stores it needs for winter by the end of July. If all these sealed stores are removed, sugar syrup has to be fed and this syrup has to be processed, ripened, stored and sealed; this is difficult for the bees to do on chilly days and nights in autumn, particularly the ripening and evaporating the excess water. It is as well to remember that unsealed stores are likely to ferment and fermenting stores produce alcohol which in turn causes dysentery in the honeybee.

b) All colonies require sampling for the adult bee diseases before the colony settles down for winter. If nosema is present, Fumidil 'B' can be fed with their winter rations; it would be pointless to feed an adequate amount and then find another gallon of syrup plus medicament has to be administered. If Acarine is present and the crop has been removed the colony can be treated without fear of tainting any honey for sale and the treatment (Bayvarol in lieu of Folbex VA) can be given during good flying weather.

c) Colonies may require to be requeened and it is better to know that the new queen is accepted and laying before clustering starts at 57°F (14°C).

Those colonies that are destined for the heather are prepared before they go with a young queen and hopefully return with a full brood box of stores and a super of surplus honey.

4.4.1.2 Requirements for successful wintering are as follows:

a) A sound and weatherproof hive.
b) 35lb (16kg) of sealed liquid stores.
c) A young fertile queen.
d) The colony to be disease free.
e) Good ventilation while excluding mice.
f) No disturbance from October to March.

These requirements will be examined to understand the importance attached to each.

4.4.1.3 A sound and weatherproof hive. This item must be self evident but it is quite surprising the tatty quarters in which colonies are expected to survive; roofs in particular seem to be often inadequate. Double walled hives have roofs that blow off and a secure method of roping or screwing them down is necessary. If a single walled hive roof has the right clearance, $^5/_{16}$ inch (8mm), between the brood box and the inside dimension of the roof it will not blow off; many do not meet this requirement. Weatherproofing means having the hive off the ground on a suitable hive stand allowing the floor board to stay as dry as possible.

4.4.1.4 Stores - 35lb (16kg) minimum. The beekeeper who has to feed his bees before March should not be keeping bees; he has not prepared them adequately for their winter hibernation.

a) In August after the crop has been removed and reduced entrances inserted, every frame has to be examined and the stores estimated. This is done by eye and by feel; it is surprising how quickly and expert one can become at this task. It is essential to know how much a full frame weighs when filled with sealed stores eg. 5lb (2.7kg) for a British Standard and 7lb (3.2kg) for a Commercial.

b) Having totted up the total stores of sealed honey in the colony, a calculation is required to know how much sugar to feed. Honey contains 80% sugar and 20% water approximately. Suppose the colony has 25lb of stores, then another 10lb is required to meet the 35lb criterion. 10lb of honey is equivalent to 8lb of sugar, the amount required to be fed in a suitable solution. It is frightening the number who keep bees and never do this examination and this simple arithmetic. See section 4.5 for details on feeding and the strength of syrup to feed. We leave it to the reader to calculate how much sugar to feed in kgs if the estimate of stores is 6kg.

60

4.4.1.5 A young queen. By observation over the years beekeepers have come to know that colonies winter better with a young queen compared with an older one. This sort of statement will be found throughout the literature but no explanation of why this should be ever seems to be forthcoming.

a) It is unlikely to be related to a quantitative problem of queen substance and the threshold amount available to each bee because the colony naturally reduces in size in the winter thereby allowing more queen substance per bee. Perhaps queen substance has other effects on colony well-being which are yet undiscovered. Young queens are likely to lay better than old queens and this could get the colony away to a better start in the late winter/early spring.

b) The other point is what is young; a queen in her 1st, 2nd year etc? Bro. Adam has always maintained that a queen lays better in her 2nd year particularly if she has not been stressed in the first year.

c) When requeening in the spring with queens bred the previous year will mean that these queens go into winter at approximately 15 months old and carry the colony through winter before being replaced the next March. This system works well. Whether the queen is regarded as old is doubtful, but if she is not replaced at 21 months old, her efficacy thereafter may certainly be expected to taper off quite rapidly.

d) Our own feeling in the matter, as a result of experience, is to ensure that queens of 24 months old do not lead a colony into winter if this can be avoided. Good queens for breeding purposes can of course be kept in nucs for much longer periods. Nevertheless, the reasons why young queens are better for wintering do seem obscure and the fact will have to be accepted until a more scientific explanation is forthcoming.

4.4.1.6 The colony to be disease free. It is vital to sample all colonies for adult bee diseases in August so that treatment may be administered if found to be necessary. Before uniting, which is a common occurrence at the end of the active season, the check for disease ensures that no loss originates from either of the two colonies. Examination for adult bee diseases now costs money if the samples are sent to the NBU at Sand Hutton, and the price per sample discourages beekeepers from using this service. Many counties have organised their own microscopy service and more individual beekeepers are doing their own (this must be good). On the other side of the coin more beekeepers are, for example, feeding Fumidil 'B' to all colonies before winter as a prophylactic against nosema. Although there are rumblings that this is unlikely to be detrimental in the long run (development of strains resistant to this antibiotic) it would in our opinion be a wrong course of action until a definitive paper has been prepared on the subject by someone with the right scientific ability.

4.4.1.7 Good ventilation while excluding mice. A colony during winter, if it metabolises 35lb (16kg) of honey, will be required to get rid of approximately 4 gallons of water. This can only be achieved by evaporation. The average rate is 5 pints/month or 3ozs/day. It is more difficult for evaporation to take place in the damp western side of UK compared with the drier eastern side. These are the facts, the best configuration for achieving this evaporation is still being debated in the bee press and still no one seems to agree on the subject. Our own method, which we have used successfully for many years after experimenting with various approaches, is as follows:

a) Heat escaping from the cluster causes the movement of air, warm moist air moves upwards and is replaced by cold dry air at the bottom.

b) All entrance blocks have 9 - $^3/_8$ inch (9mm) diameter holes drilled in them spaced equidistant apart across the length of the block. This gives a total cross sectional area of c. 1 square inch (625mm^2) to limit the air flow. The block turned through 90° is a normal reduced entrance block with a 4 inch (100mm) wide slot. The $^3/_8$ inch (9mm) diameter holes form the mouse guard and are 'kinder' to the pollen collectors in the spring.

c) The crown board is raised about $^1/_8$ inch (3mm) with matchsticks at each corner. This gives an exit area for air to escape of c. 9 square inches (c.5500mm^2). The feed hole(s) are covered so that the flow of air is round the outside of the cluster and avoiding the chimney effect directly above the cluster. The roof ventilators now play no part in the ventilation system.

d) The mouse guards are put in usually in September before the ivy flow and the crown board is raised as late as possible to stop the gap being propolised. It is interesting to note that if there is no ventilation at the top of the colony during the winter, then in the spring the top bars are wet with condensation and mould covers the outside combs. One associated problem is that some of the stored pollen also develops mould and is then useless to the bees. Strains of bee that produce a lot of propolis will get themselves into this situation if crown boards are raised too early.

4.4.1.8 No disturbance during the winter period. Once the colony has settled down for winter it should be left undisturbed until the following spring. Experiments have been conducted and it has been found that the cluster temperature is raised quite a considerable amount (up to 10°F or 5.5°C) by say just taking the roof off. Such increases in temperature shorten the life of the winter bee and this manifests itself in spring just when the colony requires all the bees it can muster. Hives should never be sited under trees where the drip of water from the branches can cause colony disturbance.

4.4.1.9 Other points of interest are:

a) The green woodpecker can spell disaster for a colony if they direct their attention to boring through the side of the hive. They are usually troublesome in very cold weather when they cannot find forage in the hard ground. There are two ways of protection:

- Surrounding the hive with chicken netting.
- Covering the hive with a plastic bag (nb this interferes with the ventilation).

b) It is desirable that a colony has stores of pollen which can be used when brood rearing starts after the winter solstice. We have never found this to be a problem but there are probably parts of UK where there is a dearth of pollen. The final topping up of pollen stores is during the ivy flow in September/October (nb. winter bees are produced by large pollen consumption).

c) Plenty of bees are required for good wintering but making massive colonies by uniting can defeat the object as shown by some experiments done by Dr. Jeffree at Aberdeen University. The old adage that bees do not freeze to death but starve to death is very relevant to the wintering problem.

d) The last thing to do is to remove the hive record card from the roof to bring the final year's records up to date and to prepare new cards for the next season.

Points to remember are:

•There is no officially recommended treatment for Acarine at the time of writing. From experience the acaricide strips, Bayvarol or Apistan used for Varroosis, provide an effective treatment for this adult bee disease. Thus the annual treatment for Varroosis provides an automatic treatment for Acarine. Apistan is no longer licensed for use in the United Kingdom though it is thought at the time of writing that a license will be granted and Apistan should be available for use in 1999.

• Treatment for Varroosis should be commenced by inserting four strips of Bayvarol into the brood chamber of a colony for six weeks after the removal of the honey crop. This treatment will kill off the mites and allow the production of healthy winter bees able to survive for 6 months. See section 5.3.

• Await the results of adult bee disease diagnosis before commencing winter preparations. These results are crucial before starting to prepare the colonies for winter ie. uniting, re-queening and feeding. Only disease free colonies should be united, never those that are being treated for disease; there is the possibility that the treatment may be unsuccessful.

• Feed all colonies for winter, treating those colonies for nosema as required with Fumidil 'B' in the feed. If after inspection a colony has 15lb of stores, it will need a further 20lb of additional stores to see it through the winter; this is equivalent to feeding the colony with 16lb of sugar (nb. ripe honey contains c. 80% sugar). It has been recommended by some sources to feed Fumidil 'B' to all colonies as a prophylactic. Many bacteria can develop resistant strains when continually subjected to antibiotic treatment. There is some doubt about this mechanism working in the case of nosema mainly because the pathogen is a protozoa and not a bacterium. Until proof is forthcoming, it would be prudent to err on the side of safety and treat only those colonies with the disease.

• Any necessary re-queening required should be completed before the end of August. The authors consider it best to avoid re-queening at this time if it is possible. If something goes wrong and a colony has to put itself to rights by raising emergency queen cells, because there is a dearth of drones at this time of the year there is less chance of a successful mating.

4.4.2 September/October: Stocks that have been to the heather will be returning during early September and these will require checking for stores and feeding if required as well as treating for Varroosis. It is seldom that colonies come back without their brood chambers adequately filled and feeding is a rarity. The final preparations for winter are to be completed before the autumn evenings start to develop a chill and a drop in temperature is evident.

• Mouse guards should be fitted early rather than late; mice are also making their winter preparations and seeking a dry warm place to hibernate.
• Ventilation requires attention, c. 4 gallons of water will be produced metabolising 35lb (16kg) of stores. Lift the crown board by inserting matchsticks at each corner and close the centre feed hole (the roof ventilators now become inoperative). Any air flow through the hive will be in through the entrance and out under the crown board and down the sides of the hive under the roof, the smallest opening at either the top or the bottom will control the rate of flow in conjunction with the internal temperature.
• If the green woodpecker is troublesome in the apiary area then protection should be added to the stocks. Polythene can be taped or tied onto the hive sides to deny a foot (or toe) hold to the bird, in which case care should be taken not to interfere with the ventilation. Chicken wire attached to

63

the hive and kept about 6 inches (150mm) off the sides denying the bird's beak access to the woodwork is also a suitable method of protection.
- The final preparation for winter is the roping down of the stocks if this is considered necessary.

4.4.3 November/December/January: Providing all the preceding preparations have been completed satisfactorily, there is nothing the beekeeper can do to assist his bees to get through the winter to the next spring. Apiaries should be inspected regularly (say once a fortnight) or after a particularly bad spell of weather to ensure all is well. The hives should not be touched, even taking the roof off will raise the temperature of the cluster unnecessarily. If something is amiss then, of course, it must be put to rights; unfortunately most mishaps seem to be associated with vandalism. Heavy snow falls can cover the hive entrances; any snow should be very quietly cleared away without alarming the colony in order to maintain the ventilation at the bottom. The practice of feeding candy on Christmas Day still seems to persist quite widely around the country. It is a nonsense and an unnecessary disturbance to the colony if it has been prepared properly for winter.

4.4.4 February/March: The work to be done will depend very much on the weather and whether the colonies are flying. In the south of England the first warm day after the 3rd week in February is the author's guide.

- The first task is to change all the floorboards in the apiary.
- Mouse guards should be removed and reduced entrance blocks put in.
- The last job is to quickly check for sealed stores, remove the matchsticks, lower the crown board and at the same time uncover the feed hole to provide some top ventilation. This all takes a couple of minutes per hive and less with two people; the bees are hardly disturbed.
- When the first task is undertaken on a colony, the hive record for the season should be started and the first entry made. The authors keep two records, one on a hive card kept in the hive roof and the other as a computer printout on a clip board which is up-dated in manuscript in the apiary and then later on the computer. The latest version from the computerised records are taken on the clip board on the next apiary visit.
- In some areas of UK where there is an inadequate supply of early pollen, stocks which are to be used on the rape are fed pollen patties to stimulate brood rearing rather earlier than would have occurred naturally; the patties are put on about the beginning of February.
- As the weather becomes warmer so the colony will start to fly and forage, with a result that stores are used up at a much greater rate. Water has to be collected to dilute the stored honey (only a 50% sugar solution can be metabolised by the honeybee) and this is a good time to start training the bees to a water supply close at hand.

4.4.5 March: If queens have been overwintered in nuclei this is the time for introduction on a warm sunny day and the bees are flying well.

- This is the first time that the colony is inspected and while removing and caging the old queen, the colony should be assessed and a sample of bees taken for testing for the adult bee diseases. Queens from the over-wintered nucs are then caged after marking and clipping if this is your style of beekeeping; in the authors' opinion both are virtually essential. If colour coding is used, a quick method of remembering the colours is as follows:

COLOUR	LAST DIGIT OF YEAR	READ DOWN
W hite	1 or 6	W hich
Y ellow	2 or 7	Y ear
R ed	3 or 8	R eared in
G reen	4 or 9	G reat
B lue	5 or 0	B ritain

The old queens are introduced in Butler cages into the nucs to keep them going until queen rearing starts again later in the year. The newly marked and clipped queens are introduced into the colonies also in Butler cages. The whole operation can be completed very quickly and the failure rate at this time of the year is very low. The advantages of this system are:

a) While the new queens are in nucs from May to March the characteristics of the queen's progeny can be assessed; only those that are suitable are used eg. good tempered bees, disease free.
b) Finding queens in March with small colonies is made much easier.
c) The nucs are virtually self-supporting having been made up at queen rearing time, the previous May.
d) Spare queens are available at any time of the year if something does go amiss.

• All the queen cages should be removed during the following two days and a quick check made to see that all is well and the queens are laying in both the nucs and the colonies. Any colonies which are short of stores should be fed.

4.4.6 April: The spring flow will start during the month.

• This is the time that regular colony inspections for swarm control should commence.
• Old comb for replacement is placed at the outer edges of the brood box ready for replacing during the 2nd or 3rd week in the month.
• If any brood chambers are to be changed for repair, maintenance and disinfection, the colony can be quickly changed to a clean box at this time of the year before supers are required. Supers are added as required above the queen excluders which go on with the first super.
• Supers are added when all except the two outside frames are covered with bees (rule of thumb for both brood box and supers). Colonies should be building up very quickly and it is better to over-super rather than under- super early in the season.
• Colonies should be selected for fruit pollination or for going to the rape, which will be coming into bloom during April.
• Any colony which is not building up or seriously lagging behind other colonies should be singled out for a special investigation to try and determine the reason. If it can be shown that it is disease free then a re-queening job is more than likely necessary.

4.4.7 May: Usually a very busy beekeeping month with stocks being brought back from the rape and from pollination contracts.

• Regular inspections are continued for swarm control and supers added as required. At every inspection of the colony the following should be checked:

a) Are there sufficient stores to last to the next inspection if there is no income available?

b) Is the queen present and is she laying normally?

c) Is there any sign of disease?

d) Is there sufficient comb space for the queen to lay and for the bees (remember many foragers will be out when the colony is inspected)?

e) Has the colony built up since the last inspection and/or are there preparations for swarming?

• Towards the end of the month queen rearing should start and arrangements made for making up any nucs that may be required.

• Removal and extraction of the spring crop may also be done during the month and will be necessary if the crop is rape to prevent granulation of the honey in the comb. On this point it is necessary to know your area well; although the stocks in the apiary may not have been moved to the rape, it is extremely attractive to bees and they will fly a long way to work it particularly if other sources of nectar are a bit sparse.

• Depending on the weather and colony size the reduced entrance blocks may be removed.

4.4.8 June: The objective is to provide the maximum foraging force and colony size by the end of this month in order to take full advantage of the main flow.

• The colony should still be expanding and further supers may be necessary. This month is notorious in UK for a dearth of nectar and known as the June gap; occasionally it does not happen (eg. in 1989 when nectar continued to flow from March through to the end of July).

• If a spring crop has been extracted, colonies may be so short of liquid stores as to require feeding. This requires extreme care to ensure that sugar syrup is not stored in the supers.

4.4.9 July: The main flow usually starts (UK south coast) during the first week of this month and this is what the beekeeper has been preparing for since last August. The colony should be at its peak population just as the flow starts. It is all over by the last week of the month.

• Swarm control inspections are required but it is unlikely more supers will be required for bees (hopefully for honey if the supers are being filled and capped). With 3 or 4 supers on the colonies it is a hard job lifting them off for swarm control and life is much easier with two people at the job.

• When the flow is complete and the crop ripe then it should be removed and extracted straight away.

• Reduced entrance blocks should be put in to discourage robbing.

• The wet supers should be returned to the hives for drying up after extraction unless it is preferred to store them wet. In a home apiary near neighbours it cannot be over-emphasised that wet supers should only be returned to the stocks after dark. Many beekeepers do not understand the reason for this and why literally hundreds of foragers will go milling round the apiary for up to c. 100 metres causing great annoyance to any neighbours in a matter of a few minutes (nb. the round dance and the aroma of honey on the outside of the supers direct from the extracting room). There is a very great possibility that robbing can be started under these conditions.

• After removing the main crop, any stocks which are scheduled for the heather must be prepared and transported (generally by the end of the month on the south coast). The essentials for the heather stock are:

a) A current year queen (re-queening may be required) to try and keep the brood production going.

b) There should be a very full brood chamber with brood on all frames.

c) The colony should have plenty of stores to see it through until the heather starts to yield.

d) There are mixed opinions on whether the stocks should be manipulated and managed while on the moor (eg. removing brood frames when they are empty to induce greater storage in the supers).

e) Drawn comb is generally necessary (usually very cold at 1000ft altitude) in the supers.

4.5 The preparation of sugar syrup and how and when to feed bees.

4.5.1 The reasons for feeding a colony sugar are shown below:

a) To provide adequate stores for winter (rapid feeding).

b) To provide emergency stores in the season between colony inspections (rapid feeding).

c) As a means of administering drugs (generally rapid feeding).

d) To stimulate the queen to lay (usually slow feeding).

e) To prevent starvation when the colony is about to succumb (rapid).

f) To enhance wax production and the drawing of foundation and comb (slow or rapid depending on circumstances, eg. a swarm on foundation is fed rapidly).

g) When a colony has an inadequate foraging force, eg. an artificial swarm which is short of stores (rapid feeding) or after spray poisoning losses.

4.5.2 The types of feed that are fed to colonies of honeybees:

a) The standard feed is white refined household-quality sugar either from cane or beet sources (ie. refined sucrose). No brown or unrefined sugar is permissible.

b) It was recommended at one time to feed candy or fondant. It is now used only for special applications eg. micro mating nucs or the like). If cream of tartar or vinegar is contained in the recipe, both are toxic to bees cf. refined sucrose. It is best not to feed either candy or fondant if it can be avoided.

c) Dry sugar (again refined sucrose) is used by some beekeepers in a tray type crown board usually in the early part of the year supposedly as an insurance policy. It is not recommended because unless water is provided it is extremely difficult for the bees to produce enough saliva to dissolve the crystals. A variation on this is to wet the sugar to assist the bees to take it.

d) Honey. This should only be fed when it comes from the beekeepers own apiary and is known to be disease free. Many imported honeys carry AFB spores and are highly dangerous and must under no circumstances be used.

e) Pollen patties are often fed in the early part of the year to provide additional protein where pollen may be in short supply or where colonies are being induced to start brood rearing early. There are two types, namely pollen substitutes (fat-free soya flour) and pollen supplements using trapped pollen; again the source of pollen should be from the beekeepers apiaries from disease free colonies.

f) A comb of sealed honey can often be usefully taken from a disease free colony and used in another requiring urgent liquid stores.

4.5.3 The precautions to take when feeding honeybee colonies:

a) There should be no spilling or dripping of syrup anywhere in the apiary.
b) Precautions should be taken to prevent robbing (reduced entrances and bee-tight hives).
c) Feed should only be administered in the evening just before dark.
d) No sugar syrup should find its way into the supers and be mixed eventually with honey for extraction and sale.
e) Only pure white refined and granulated sugar should be used.

4.5.4 Preparing syrup for feeding: Generally there are two types of mix, a thick syrup for autumn feeding which will be stored more or less immediately and thin syrup for spring or stimulative feeding which is to be consumed without storing. Most of the past literature quotes the following:

Thick	2lb sugar to 1pint of water gives 61.5% sugar concentration.
Thin	1lb sugar to 2pint of water gives 28.0% sugar concentration.
Medium	1kg sugar to 1 litre of water gives 50.0% sugar concentration.

Since the bee requires a concentration of 50% for it to digest and metabolise the sugar then it is clear which is the best one to use if they are to use it straight away. If sugar syrup is to be mixed with cold water, it will be found difficult to obtain a complete mix with 2lb to 1 pint. The authors use a mix with cold water of 7lb to 5pt. in an old washing machine (top loader with central agitator). The concentration works out to be 52.8%, less than 61.5% and hence giving the bees a bit more work to do ripening it to 80% for storing and sealing. As we feed for winter immediately after extracting in August, this causes the bees no distress as they have plenty of time to get their larder in the order they require it before the cold nights set in.

4.5.5 The most common types of feeders in use.

The requirements of a good feeder are to allow the bees to take the syrup at the rate required by the beekeeper for the management of the colony, while at the same time preventing the bees from drowning in the syrup. Finally, when the feeding is finished, access should be provided for the bees into the feeder so that the bees can clean and dry it up (a job they can do very efficiently given the chance). There is quite an array of feeders available, not all of them meeting the criteria above and many of them being manufactured in materials that can corrode or are difficult to clean. A further disadvantage of some types is that they are capable of being propolised by the bees so that without maintenance they become unusable. The various types commonly available are listed below:

a) Contact feeders: these come in a variety of shapes and sizes but are all similar in design having a container with a close fitting lid. The lid has a series of small holes or a small piece of wire gauze through which the bees take the syrup when it is turned upside down over the feed hole or directly onto the frames in the colony. The number of holes regulate the speed that the bees can take the contents. It has the advantage of being cheap and can be readily made at short notice from a bewildering assortment of household containers. The disadvantages are as follows:

68

1) The bees quickly propolise the small feed holes as soon as it is empty.

2) As the contents are coming to an end, a change in temperature can force the last remaining content out causing a minor flood of syrup in the hive (usually cleaned up quickly by the bees).

3) They are a bit messy to fill and invert without spilling syrup unless one is very careful.

4) An eke (a super or brood box without its frames) is required in order to house the feeder under the roof.

b) Round top feeders: are very widely used in UK and are intended to be placed over a feed hole in the crown board. The capacity varies from c.1 pint to 2 or 3 pints depending on the diameter. The height is usually about $3\frac{1}{2}$ inches (90mm). The entry is via a tube in the centre and down the outside of the tube to the syrup. The whole of the centre feeding area is enclosed by a removable cover for cleaning. Older versions were made of metal but now most are manufactured in plastic which is better from a corrosion point of view. This type of feeder is easily filled in situ without the bees escaping in the process. Again an eke is necessary.

c) Miller feeders: were designed by Dr.C.C.Miller in USA and consist of a tray about 3 inch (75mm) deep with dimensions in the horizontal plane exactly matching the external sizes of the brood chamber or supers of the hive it is intended to fit. The entry for the bees is via a slot in the centre extending from one side to the other; again it is provided with a cover to prevent the bees from escaping. The capacity is from 1 to 2 gallons. It allows many bees to feed simultaneously thereby allowing very rapid consumption of the syrup (a strong colony can finish the contents in 24hrs.). Construction is generally in wood with all joints glued to prevent leakage. For bottom bee space hives, a bee space is required on the under side of the feeder.

d) Ashforth feeders: are virtually identical with the Miller feeder except that the feeding slot is placed at one side allowing the hive to be tilted slightly, thereby permitting all the syrup to flow towards the feed slot which is impossible with the Miller type and therefore an improvement. The advantage of allowing all the syrup to be consumed before the tray is opened to the bees for cleaning is that there are no pools of syrup for the bees to drown in.

e) Bro. Adam feeders: are similar to the Miller and Ashforth except that they have a central entry similar to the Round top type feeders. They are becoming more popular in UK due to some equipment suppliers now manufacturing them. The feeders on the stocks at Buckfast Abbey double up as a crown board (therefore every stock has its own feeder).

All the above type feeders are designed for top feeding. Other feeders are available for internal feeding and bottom feeding (which is seldom practised in UK). The internal feeder is in the form of a brood frame with wooden sides and an opening at the top to allow access to the bees. The frame feeder is used for feeding nucs; the capacity is inadequate for a colony and few would wish to open the colony in order to feed it. The bees propolise the float arrangement in the frame feeder, the float prevents the bees from drowning, a fixed float is useless and a weakness in this design of feeder.

4.5.6 The amounts of food to be fed.

4.5.6.1 Emergency feeding. It is necessary to know the amount of food that a colony requires during the season so that, after an inspection, the beekeeper can determine whether it will require feeding or whether it has sufficient stores to the next inspection. The worst case must always be considered and that is when the colony sends out its foragers and they are unrewarded in their search for food.

A flying bee uses 10mg honey per hour while foraging for an average time of 5 hours per day. If the colony has 13,000 foragers ($^1/_3$ of the total population) and the next inspection is 7 days away, then the colony should have 10lb of liquid stores ie. 10lb = $(13000 \times 10 \times 10^{-3} \times 5 \times 7) \div 454$

Therefore, if the colony has less than 10lb of stores it may require emergency feeding if there is no income and the weather is inclement. The amount required is likely to be small, ie. a few pounds. The same considerations are applicable to nuclei and many a nuc has died out due to starvation because of ignorance of the beekeeper not understanding the little colony's food requirements.

4.5.6.2 Winter feeding. We are alarmed and distressed by the large number of beekeepers who either don't know how much food a colony requires for winter or, if they do know, have no idea how to calculate how much it should be fed. The losses each year in the UK due to starvation amount to many thousands of colonies according to a MAFF survey some years ago. We doubt if the situation has changed. If the RSPCA knew more about bees they would be taking some action against the offending beekeepers.

The calculation is a simple bit of arithmetic and the starting point is a colony inspection in August. Each frame in the brood chamber is inspected and the amount of liquid stores estimated on the basis that a BS frame when full and sealed with honey weighs 5lb. A Commercial frame holds 7lb.

A strong colony requires c. 35lb to see it through to the spring without feeding early in the new year when stores are used up very quickly. It is often said that a beekeeper who has to feed his colonies in the spring should not be keeping bees! To illustrate the simplicity of the calculation, assume the colony has 25lb of stores after the inspection. The colony requires 35 - 25 = 10lb of additional stores or the honey equivalent thereof. How much sugar must be fed in syrup form to provide the equivalent of 10lb of honey? 1lb of honey contains c. 0.8lb of sugar, therefore, 8lb of sugar should be fed in syrup form. If the colony required 15lb of additional stores then the amount of sugar = 15 × 0.8 = 12lb sugar. It is as simple as that and yet very few beekeepers take the trouble to do the job properly and many colonies starve to death.

4.5.6.3 In the case of starvation. Hopefully this will never occur but the signs are unmistakeable; there is usually a pile of immobile bees on the floor board and if there are bees on the combs they will be very immobile also, just about clinging on. Warm 50:50 syrup sprayed onto the bees is the first aid treatment to allow them to get their tongues immediately to a syrup which can be metabolised immediately with out recourse to water or ripening. If you are in time the whole colony will slowly come back to life and return to the combs; it doesn't take long. Further light spraying before equipping the stock with a feeder and a gallon of warm syrup of the same strength. Don't let it ever happen again.

4.5.7 The timing of feeding a colony of honeybees.

All feeding of colonies of honeybees should be undertaken only in the evening when it is just getting dark.

The reason for this is not explained at all well in most books on bee husbandry. This is curious because it is so important particularly when bees are being kept in gardens at close quarters with neighbours.

The reason is that as soon as food is given to a colony during daylight hours, the scout bees will be alerted and will start roaming the immediate neighbourhood for the source. It seems to be a shortcoming of the communication system of the bees. Presumably a round dance occurs and out go the foragers to seek the source and mayhem starts in the apiary with the attendant possibility of robbing being started also. It seems that the colony has no sure means of indicating to the other foragers in the colony that the source is just above them over the brood chamber in a feeder.

Bees are not equipped for night flying and will not fly in the dark. Hence, all feeding should be done at night. The same goes for putting wet supers back on a colony for drying up after extraction.

4.5.8 Other points relating to feeders and feeding:

• Each hive should have its own feeder. When feeding starts, particularly in the autumn, all stocks should be fed at the same time.
• There are advantages in combining the feeder as the permanent crown board; it is always available for use and if it stays on one stock it cannot pass on disease by using it on another colony.
• Open tray feeders with straw or polystyrene chips floating in the syrup are often messy and not particularly efficient, the bees often seem to find the 'deep end' and drown in the syrup. Not recommended.
• It is good practice to check the feeders each year for leaks with water before being brought into use.
• Communal feeding has been advocated by some authors by providing a common feeder in the apiary for all colonies to fly to and help themselves. We do not recommend it as the disadvantages far out weigh the advantages. No control can be exercised over what each stock needs and takes. Disease can be spread by this means and you are likely to be feeding someone else's bees!

4.6 The need to add supers and the timing of the operation.

4.6.1 Definitions:

• Super. A box containing frames/combs placed above the brood chamber for the eventual storage of honey. The word 'super' is derived from the Latin word super meaning above (eg. super-script as opposed to sub-script). Supers are generally shallower than brood chambers because of the weight when full of honey; other than this, there is no technical reason why they shouldn't be any depth providing the frames can be accommodated in the extractor.

• Supering: is the process of adding supers to a colony above the brood chamber either with or without a queen excluder under the super(s).

• Top supering: is the term given to adding further supers to a colony but always adding them on top of any existing supers.

• Bottom supering: no prizes for guessing that the supers are added at the bottom of the pile and always next to the brood chamber.

4.6.2 Principles involved.

During the spring build up the annual colony population cycle (see Appendix 3) shows a very rapid increase in adult bee population from the beginning of March. It is not long, providing the weather is fine, before the brood chamber starts to fill up with both brood and bees and if nothing is done there will be insufficient room for the emerging brood. Additional space is therefore provided by adding supers, usually one at a time, as required by the colony build-up. On this basis, supers are for bees and, indeed, this can be very true if the colony is using most of or all its income. In such a situation nothing will be stored in the supers and it will be used solely as a parking place for bees in the colony. If this additional space is not provided, overcrowding will occur and this congestion in the hive leads to a breakdown in the food sharing pattern and subsequent distribution of queen substance with a result that the liability to swarm is greatly enhanced.

Reference to section 3.7, where the manipulation and ripening of nectar to honey was discussed, it will be appreciated that large areas of comb are required for the nectar/honey to be 'hung up' to dry in order to evaporate the water. The change in volume of nectar (30% sugar concentration by weight) to honey (80% sugar by weight) is approximately 100:30 thus requiring c. 3.3 times the space for nectar compared with the space required by the finished product.

The calculation looks like this:
  1 litre honey weighs c. 1400g (80% sugar 20% water by weight)
  1 litre of water weighs 1000g

  1400g honey = 1120g sugar + 280g water and
  1000ml honey = 720ml sugar + 280ml water
  Therefore 1g sugar has a volume of $720 \div 1120 = 0.64$ml

CONCENTRATION	SUGAR	WATER	TOTAL
Nectar 30%	30g	70g	100g
	19.2ml	70ml	89.2ml
Honey 80%	30g	7.5g	37.5g
	19.2ml	7.5ml	26.7ml

It will be seen that 89.2ml of nectar (30%) when processed to honey only requires a volume of 26.7ml; this is a change of $89.2 \div 26.7 = 3.3$.

There are only two principles involved as detailed above and summarised below:

  1) Primarily to provide space for bees, and
  2) to provide comb area for ripening nectar.

If adequate space is provided for evaporation then it will be clear that there will be adequate space for honey storage.

4.6.3 Other points related to supering:

a) By experience it has been found that a good working guide for supering is to add a super when the bees are covering all but the two outside frames of the top box or initially the brood chamber.

b) It is better to super early in the spring and be somewhat tardy about adding supers in July when the main flow is on unless this is absolutely necessary.

c) In general, top supering is the most widely used method of adding supers. Bottom supering is advantageous if the frames in the super contain only foundation; ie. placing them above the brood chamber, the warmest place in the hive for the wax-makers to work.

d) There are quite a few beekeepers that super without the use of a queen excluder; however, the majority use an excluder. Again there are beekeepers who advocate not using an excluder in the spring when the first super goes on to encourage the bees into it more quickly. It is true the bees always seem to be somewhat tardy about occupying the first super but this may be due to observing the rule of being just a little ahead of the bees requirements (super when the two outside frames are uncovered).

e) See section 2.5 on spacing of frames. Narrow spacing is essential with foundation, later the spacing can be widened out to 2 inch (50mm) when the foundation is drawn and being filled with honey. 2 inch (50mm) is the maximum width of comb that a wild colony will build for the storage of honey. Cut comb containers have been designed on this thickness.

f) If the super contains frames with foundation only, it is best to provide one or two frames of drawn comb in the middle to encourage the bees into the super more quickly.

g) Wet stored supers are more attractive to the bees in the spring cf. dry supers.

h) If the supers are added too late, the expanding colony of bees will build brace comb in all available spaces in order to store the honey. This will cause extra work for the beekeeper at the next inspection.

4.6.4 The importance of supering as a factor in swarm prevention.

The most important factor which causes a colony of bees to swarm is the lack of an adequate threshold level of queen substance throughout the colony which was discovered and proved by a series of experiments by Dr.C.Butler who worked as a scientist at Rothamsted. However, it is well known by observation, but not proved, that other factors have an influence. These other factors include:

1. season
2. shade
3. state of flow
4. strain of bee
5. manipulations
6. weather
7. ventilation
8. district
9. comb space (queen)
10. comb space (honey)

Considering the two principles of supering, it will be clear that by providing additional empty comb and thereby additional space, not only are the last two conditions relieved but ventilation is also improved. The additional comb space in the supers provides the needed storage space for nectar and honey leaving the comb in the brood chamber for the queen to lay in. The over-riding factor is the prevention of congestion within the hive and the efficient distribution of queen substance.

4.7 An elementary account of one method of swarm control.

4.7.1 General. The subject of swarm prevention (which logically should come first) and swarm control is so vast and so important that to confine the discussion to only one method of control would defeat the object of having a reasonable understanding of the subject for examination purposes. Firstly, it is necessary to clearly understand the difference between prevention and control and secondly to be able to detect the preparations for swarming and to know how often to inspect the colony in order to detect the preparations.

• Swarm prevention: is the action(s) taken by the beekeeper to prevent the colony reaching the state whereby it starts to build queen cells and preparing to swarm.

• Swarm control: is the action(s) taken by the beekeeper to thwart the colony in its endeavours to swarm once the preparations for swarming have been started thereby preventing the loss of bees.

• Detection of swarming preparations: this is necessary before any swarm control measures are put into practice. The ability to detect the preparations is prerequisite to any control actions.

• Frequency of inspections: for swarm detection is dependent on whether the queen is clipped or unclipped. It is beyond the scope of this syllabus to prove that:

When the queen is unclipped the colony must be inspected every 7 days, and
when the queen is clipped the colony may be inspected every 9 days.

• It is necessary to control swarming for a number of reasons which are frequently overlooked by many beekeepers, these are:

a) A colony that swarms is unlikely to produce a surplus of honey cf. the colony that does not lose its bees; this is to the detriment of the beekeeper but of little consequence to anyone else.
b) Most of the general public are petrified of bees and if not petrified then they have an innate 'api-phobia'. In an urban environment it is essential that no swarm settles on a neighbouring property (this unfortunately cannot be guaranteed).
c) When a colony swarms, there are many thousands of bees flying around which most people find very frightening and can be classed as a nuisance in an urban or suburban environment.

4.7.2 Swarm prevention. In section 3.9 the role of queen substance in relation to swarming was discussed and in section 4.6 the importance of supering as a factor in the prevention of swarming was examined. It is important to understand the role of queen substance and the inter-relationship between food sharing and congestion in the colony as the trigger in the process of swarming. The prerequisite in swarm prevention is that the colony must be headed by a young queen in order that each bee in the colony is assured of its minimum threshold quota of queen substance. When an adequate supply is available at its source (ie. the queen), the next most important factor in swarm prevention is to ensure that the supply can be distributed around the colony; this can only happen if there is plenty of comb for the bees and hence no congestion. Add good hive ventilation and the beekeeper can do little else in the way of prevention. Nevertheless, having done all this a colony may proceed to build queen cells and it is incumbent on the beekeeper to control the issue of a swarm.

4.7.3 Detection of swarming preparations. It is very important for every beekeeper to be able to recognise the preparations for swarming while undertaking a routine 7 or 9 day inspection. At the beginning of the season the colony will have no drones and no queen cups (easily recognised, being almost identical in shape and size to acorn cups). As the colony builds up drones will appear and queen cups (known in some parts of the country as play cells; reason unknown) will be built around the outer limits of the brood nest. It is important to examine them closely. If eggs are found in them it does not follow they will be turned into queen cells; in many cases the eggs are removed or eaten by the bees. However, if the cup contains royal jelly, a larva will also be present which is sometimes difficult to see floating in the pool of liquid as the egg may have only recently hatched. This is the sign that preparations for swarming have commenced and swarm control proceedings must be initiated. The simple rules are:

a) Dry queen cups (nothing in them or egg only); the situation can be left to the next inspection.
b) Charged queen cups (containing royal jelly); initiate swarm control procedures.
c) Half-built charged queen cells; ditto, initiate swarm control procedures.
d) If one or more sealed queen cells are present, the chances are that the colony has swarmed and an inspection date was missed or signs of swarming were missed during the last inspection.
e) If all the queen cups have a dull matt finish on the inside, preparations for swarming have definitely not started; the cells will be polished before the queen will lay in them.

4.7.4 The frequency of inspections.

The timing of inspections for swarm preparation as stated previously is every 7 days for a colony with an unclipped queen and every 9 days with a colony with a clipped queen. It is necessary to understand the mechanism involved and the process of events inside the colony. Subject to the weather being favourable, a swarm will issue with the old queen just after the first queen cell is sealed. If the weather is inclement, then the swarm will not emerge until the weather has improved. The swarm can contain the old queen plus virgin(s) if the time is 8 days after the first queen cell was sealed or rarely virgins only (the old queen having been killed by the virgin(s)) if the time is 8 days or more.

The first swarm, known as a prime swarm, generally leaves the hive with the old queen and about 50% of the workers in the colony. This is a huge loss of honey gathering potential (see Appendix 3). The old parent colony now contains the remaining 50% of the work force, plus brood and sealed queen cells. Generally, after about 8 days a second swarm, known as the 1st cast will issue with about 50% of the remaining work force with one or more virgin queens. The original colony now has only 25% of its original strength and has become a useless unit from a honey harvest point of view.

75

It will now be clear that it is vital to keep rigorously to the inspection dates so that a method of control can be effected to ensure that no bees are lost by swarming.

4.7.5 Swarm control. Over the years there have been three major theories of swarming, namely:

- Brood food (postulated by Gerstung in 1890).
- Congestion (postulated by Demuth in 1921).
- Queen substance (postulated and proved by Butler in 1953).

Only the latter satisfactorily explains why a colony swarms and is now accepted as the only correct theory of swarming. Congestion prevents queen substance from being distributed around the colony and is therefore, in itself, not a theory. The brood food theory was accepted for a long time but is now regarded as being incorrect; it is based on the surmise that as the colony builds up, an excess of brood food is produced and this is used in queen cells that are built to absorb this surplus.

Most swarm control methods involve finding the queen and some require finding and destroying queen cells which in turn requires shaking bees off frames. Allied with these operations of controlling, regular inspections are required to know when to undertake them. Such inspections and control can only be undertaken with good tempered bees and ensuring the 'right strain' is a necessary part of swarm control. When the colony becomes bad tempered, regular inspections get abandoned, the colony swarms and the bad temper is promulgated further around the district. This indeed must be classed as anti-social behaviour on the part of the beekeeper. The authors believe that a major contributory cause of such situations arising is the present day obsession to wear gloves to manipulate the colony. If the norm were no gloves (kept in reserve for the real emergency) then colonies would be requeened before situations got out of hand. If colonies cannot be handled without gloves, then the handling technique or the strain of bee is at fault; both should and can be corrected without delay. Anyone keeping bees in an urban garden should consider this point long and hard.

Many swarm control methods involve the use of double brood boxes (eg. Snelgrove, Demaree, etc.); it is not proposed to discuss these here, but to confine the discussion to one method requiring additional equipment, ie. the 'Artificial Swarm'. If this method is thoroughly understood it can be used for the whole of one's beekeeping career. The method provides a high degree of control and has a success rate approaching 100%. It does require finding the queen; it is infinitely easier to find the queen if she is marked and marking is regarded as a high priority for effective swarm control.

4.7.6 The Artificial Swarm. This method must become common knowledge to anyone keeping bees. Briefly, when the operation has been completed the queen and one frame of bees plus empty comb to fill a new brood chamber remain on the original site and the colony with all the queen cells and remaining bees is put on a new site. All foraging bees return to the original site and, with the queen, form the artificial swarm. The old colony with only house bees and queen cells rear a new queen without swarming. This is the basis of the method, other more detailed points of interest are as follows:

76

a) A new hive is required, ie. floor, brood box and frames of comb or foundation, crown board, feeder and roof. This will all end up on the site of the parent colony (the one that is going to swarm).

b) Assemble the new hive behind the parent colony without its roof or crown board. One frame should be missing in the middle.

c) Quietly open up the parent stock using as little smoke as possible and examine very carefully and very slowly every comb until the queen is found.

d) Check there are no queen cells present on this frame with the queen and place it into the space left in the middle of the new hive. Close up the frames and replace the crown board.

e)Fill up the brood chamber of the parent hive with an additional frame and close up the hive. Then move it to a new position say about 6 feet away from the original position.

f) Move the new hive with one frame of bees and the queen forward into the original position and the job is virtually complete.

g) If the parent colony has supers, then where should these end up; on the artificial swarm with the foragers or on the original stock with the queen cells? Most books show the supers on top of the artificial swarm on the original site. The old stock (now weakened by c.50% of the total original number of bees) on a new site may need feeding and could be robbed. It seems logical to put the supers on the old stock and feed the artificial swarm which in all probability will have foundation to pull out and also, doing it this way, there will be no possibility of contaminating the supers with sugar syrup.

h) Put a feeder on the artificial swarm and feed it a gallon of syrup (50:50) after dark when the divided stocks have settled down. The job is now complete.

Other points:

• Many books recommend moving the original stock a second time, to the other side of the artificial swarm, to draw off any additional foragers 7 days after the manipulation and before a virgin has emerged. Unfortunately the rationale behind such a move is not explained. It does provide additional foragers for the artificial swarm but is not essential to the success of the manipulation.

• It is unnecessary to destroy all but one queen cell in the original stock as the removal of foragers reduces drastically the strength of the colony and the bees will undertake the destruction themselves.

• If necessary, the operation can be completed on the same site with the artificial swarm below and the old stock on top above the supers plus a swarm board or similar. In this instance the entrance to the top brood box should be at the rear. If it is done this way, then any feeding will be confined to the top stock.

• The advantage of this method of swarm control is that it is virtually 100% successful and can be performed on any stock. Additionally, brood rearing continues with the old queen and the two units can be united at a time suitable to the beekeeper. The disadvantage is that additional equipment is required.

4.8 How to take a honeybee swarm and how to hive it.

4.8.1 General considerations about swarms. Before considering how to take and hive a swarm, a few points of interest are listed below which will assist in understanding the task to be undertaken:

a) There are two types of swarm, a prime swarm and a cast. They differ in size, the prime swarm containing about 50% of the original colony and the casts being very much smaller (as little as a cupful of bees in some cases).

b) Swarms settle initially within a few metres of the original colony. The prime swarm is generally predictable in its behaviour, remaining where it first settled until it has decided on a new nesting place before moving (a matter of a few hours or rarely a over a week). In very rare cases they never make up their 'mind' and try to establish comb and a nest outside where they have settled; they invariably perish with the onset of cold and bad weather. Conversely, a cast is very fickle and will take off quickly (can be within the hour) and resettle somewhere else close at hand or at a distance.

c) The settling place can be almost anywhere; on a post, on a wall, in a tree or bush, on a fence, under eaves, high or low, etc. Because of this diversity, only broad guidelines can be enunciated for taking swarms and the beekeeper will have to use his own ingenuity, depending on the situation.

d) Swarms when they first emerge are generally very docile (even bees of doubtful temper) because they have gorged themselves full of honey before departure for immediate future comb building operations. The longer the swarm hangs up after its emergence, the more its behaviour will return to the normal temperament of the bees and this can be anything from good to aggressive. For this reason swarms from unknown sources should be approached with caution and every effort made to determine their history (eg. how long has the swarm been there?).

e) After the swarm has settled it forms a cluster with an outside shell of bees about 3 inch (75mm) thick with a hollow centre. The outer shell has a small entry/exit hole about 1inch (25mm) diameter. Close examination of the outer surface will reveal, after about an hour, dancing bees. There may be dances indicating different locations while the swarm is 'arguing' which site to choose as a final nesting place.

f) Swarms can vary very considerably in size. The weight of the swarm can range from a few ounces for a cast to 8 or 10lb for a prime swarm. The skep or box to transport the swarm must be capable of carrying the load.

4.8.2 Taking a swarm. There are different methods depending on the situation of the swarm; these can be classified into the following broad categories:

a) Shaking the swarm into a skep.
b) Smoking the swarm up into a skep.
c) Enticing the swarm into a nuc with a chemical swarm lure.
d) Using a frame of brood to attract the swarm.
e) Brushing the whole swarm down into a more convenient place so that it can walk into a skep.

The prerequisite of taking any swarm is to get the queen into the skep. Once this is done all the rest of the bees will join her.

The essential equipment required for taking swarms is as follows:

a) A good sized skep with a mouth of about 14 inch (350mm) diameter or larger. Some swarms are often quite wide and anything smaller makes the operation that much more difficult.
b) A second small skep of 9-10 inch (c.250mm) diameter. This is useful for collecting any stragglers if the first shake is not as clean as it might be.
c) A piece of cloth or net curtain to place under the skep/ box which will be used to close the mouth of the large skep by gathering it up and tying over the top of the skep.
d) Secateurs, string and small block of wood to put under the skep (there never seems to be a suitable stone in sight at the right time).
e) Butler cage for caging the queen if she is found.

All the above can be kept in the skeps and ready for immediate use (maybe in the back of the car during April to June, the swarming season in UK). Additional items are:

• Smoker, fuel, matches, hive tool and veil.
• Steps and/or ladder.

Many beekeepers take swarms not because they want them but to provide a service to the community. Prompt efficient action to a call is not only appreciated by the person concerned but it makes for good public relations and enhances the general image of beekeepers.

4.8.3 Different methods of taking a swarm. Always advise members of the public to keep well back, especially if there are any children present. If the swarm is on the side of a house it would be wise to ask for all the windows to be closed.

4.8.3.1 Shaking directly into a skep. Clear away, with the secateurs, any small foliage to allow the skep to be brought up and under the swarm as close to it as possible. One sharp jerk of the main branch that the swarm is clustering on should get 99% in the skep in one go. The preparation before shaking pays dividends. Slowly turn the skep over and place it in the middle of the sheet on the ground below where the swarm was clustering propped up on one side with a small block or stone to allow the bees to get in and out. In about 20 minutes all the bees will be in the skep and foraging is likely to be starting. Leave until the evening when all the bees have returned, remove the block, gather up the sheet round the skep, tie off and carry away for hiving. If the queen has been missed in the shaking process the swarm will start coming out of the skep and resettling with the queen, more than likely in the same spot. The small skep is useful now for a second shake if this happens. Throw/shake these bees into the large skep when it is turned over, momentarily, the right way up. It is always best to wait about 20 minutes to see that all is well before departing until the evening. Collecting swarms is much easier with two beekeepers especially if steps or a ladder is involved.

4.8.3.2 Smoking upwards. Quite often the situation arises where the bees cannot be shaken off (eg. on a wall or a rugged post) and they can then be smoked up into a box or skep. Bees will always walk upwards into a darkened space. When a swarm is on a wall or flat surface this is the only time a cardboard box is better than a skep (the long flat side can be laid against the wall above the swarm). Remember to push some slivers of wood through the box before collecting the swarm, the bees cannot cling to a smooth surface, the swarm will regroup/cluster around the wooden bars. The box is brought in contact with the swarm and it is gently smoked to get them marching in. Once in they are put on the sheet on the ground as above and left to fly until the evening. Smoking a swarm upwards is a much slower operation than shaking and if the box/skep can be temporarily be fixed in position it will be a lot easier than holding it for half an hour.

4.8.3.3 Using a frame of brood. This always works in a difficult situation providing the frame can be brought in contact with the swarm. The bees soon cover it and can be shaken into a nuc box and then the frame can go back to the swarm for more bees. If the queen is seen, then the Butler cage will come into its own; with the queen inside the cage and in the nuc box or skep the bees will follow with no prompting. The queen can be released later when the swarm is hived.

4.8.3.4 Chemical swarm lure. The authors have only tried this once on a cast on a rose bush close to the ground. Two frames of drawn comb were placed in a nuc which was placed with its entrance close to the swarm. The entrance inside the nuc was treated with about $1/2$ inch (12mm) of swarm lure (French brand; in a tube to be squeezed out for use like toothpaste). The idea was to come back in the evening to collect but in minutes the whole swarm was inside the box. Her ladyship, whose garden we were in, thought it magic - and so did we! The experiment seems to have some merit for future use and adaptation for taking swarms in difficult positions.

4.8.4 Hiving a swarm. An interesting phenomenon about a swarm is its loss of 'memory' of its old nest or hive more or less immediately it has emerged and settled. The swarm can be taken straight away, hived anywhere in the same apiary and the foragers will not return to the old site. It is analogous to erasing the information on a computer disc. Why such memory erasure should take place during swarming is unknown. Bees have a memory of their original site lasting about 2 weeks when they are not in the swarming mode. There are two basic methods of hiving a swarm:

  a) Swarm board.
  b) Shaking into an eke.

The first is fun and amuses both the beekeeper and any spectators. It allows inspection of the swarm and the queen(s) present as they march in. The second method is for the beekeeper who has used the first method so many times that he no longer finds it amusing; especially if time is at a premium it is quick and efficient. Swarms are in just the right state for drawing foundation and building comb and such an opportunity should not be missed. If possible always put a swarm onto foundation except maybe for one drawn comb.

4.8.4.1 Swarm board: is so called because a board (c. 2ft. square) is placed in front of the hive sloping up from the ground to the hive entrance, covered with the sheet from the skep and the swarm shaken on to it. Bees always walk upwards and in a few minutes there is a steady procession walking up into the hive. If the bees are reluctant to start a few taps on the board with a pencil or hive tool will start the proceedings. It takes about half an hour. The brood chamber that the swarm will occupy should have one drawn comb if the swarm is of unknown origin; if eggs are found in this comb 2 days later it indicates an old queen, if there are no eggs it is likely to be a virgin and it should then be left for about two weeks before inspecting again. A feeder with 1 gallon of syrup should be provided straight away.

4.8.4.2 Shaking into an eke. A shallow eke is placed on the floorboard with the brood chamber complete with frames of foundation over the eke. The entrance should be closed with an entrance block turned through 90°. The brood box is lifted off and the swarm shaken into the eke and the brood box replaced immediately with a feeder with 1 gallon of syrup over. The swarm will walk up into the frames and 10 minutes later the entrance block can be removed (its only purpose was to stop the bees spilling out of the front entrance). The eke should be removed the next day.

4.8.5 Other points.

a) Take a sample of the swarm to check for adult bee diseases.

b) When brood is starting to be produced examine it very carefully for brood diseases. Swarms from unknown origins can be a liability.

c) If any bad temper is present when the colony has settled down, take action to requeen it immediately. Lots of stray swarms consist of bad tempered bees because their owners could not handle them, so they are a liability on this score also.

d) Two or three swarms can all be shaken into one hive at the same time if several swarms are collected on one day. The queens will sort themselves out and there will be no fighting amongst the bees.

e) Casts should be hived with a frame of brood from another colony to prevent them absconding. If more than one queen is present then early in the following morning the dead vanquished queen/s will be seen by the entrance.

4.9 The signs of a queenless colony.

4.9.1 The signs (not symptoms) of queenlessness in a colony can be readily seen by the observant beekeeper both outside the hive and also inside. The beekeeper with only a little experience can generally tell, by looking at the entrance and studying the bees' behaviour patterns, whether the colony is normal or abnormal. It is a sound practice to have 2/3 stocks in the garden and to study the entrances for a few minutes 2 or 3 times a day while the bees are out, both in the summer and in the winter through to spring. With practice it is generally possible to say whether a colony is normal or abnormal by observing the entrance without recourse to examining the interior; this is particularly useful in the early spring as winter is drawing to a close. The art should be developed by all beekeepers. Colonies seldom become queenless of their own accord, it is usually due to an error or series of errors on the part of the beekeeper and his manipulations which cause this condition. A very small percentage of colonies become queenless during the winter due to the death of the queen at a time when the bees do not have the ability to produce a replacement.

4.9.2 The signs of queenlessness are as follows:

a) If a queen is removed from a colony, within about 10 to 15 minutes, signs of queenlessness are likely to be observed at the entrance. Bees are wandering around the outside of the hive 'looking for' the queen. These bees crawl up the front and sides of the hive and appear to be in an agitated state. Conversely, if the queen is returned to the colony, it takes about the same time for it to return to normality (see Appendix 2 re distribution of queen substance).

81

b) If the colony has been queenless for some time (say 24 hours or more) the foraging will be greatly reduced, there will be apathy among the workers, some of which will be running around somewhat aimlessly.

c) The colony will become more aggressive than usual and difficult to handle.

d) The first signs inside the colony are no eggs and eventually no brood and the possible appearance of emergency QCs built over worker brood.

e) If the colony has no means of re-queening itself it is said to be 'hopeless', for example, when there is only sealed brood or no brood at all in the colony.

f) The colony starts to dwindle in size, and if it is left queenless for 3 weeks or more laying workers will start to appear and eggs will be found in an erratic laying pattern. More than one egg per cell is common and drone brood in worker cells start to appear. At this stage the colony has a 'dispirited air' (difficult to describe in words).

4.9.3 Confirmation of queenlessness. A colony, of course, should not be allowed to get to the stage of laying workers. Before this stage, a virtually infallible test is to put a frame of eggs and brood of all ages into the queenless colony. If there is no queen (either fertile or virgin) then queen cells will be started on the introduced larvae within 24 hours. If this test is undertaken too late (ie. laying workers present), it may not work. There is only one situation when this test comb may not give a reliable indication and that is after a colony has just swarmed and has a young mated queen. QCs may be built still under the swarming urge. This is not a very common occurrence.

4.10 The signs of laying workers and of a drone laying queen.

4.10.1 The signs (detection) of laying workers and description of why they occur.

4.10.1.1 Detection of laying workers:

 a) Drones in worker cells (typical raised domes).
 b) Drones produced in this way are small and abnormal (stunted).
 c) Laying pattern is scattered and haphazard (cf. drone laying queen which is compact and orderly).
 d) Colony endeavours to build charged queen cells (note: this can happen with drone laying queen but is unusual).
 e) Workers generally lay more than one egg/cell.

4.10.1.2 In the absence of the queen and brood there is an absence of pheromones from the queen and from the brood. These pheromones, particularly that produced by the queen, inhibit development of a worker's ovaries. Workers' ovaries develop in the absence of these pheromones and some workers start laying after about 21 days in the queenless state.

4.10.1.3 Cause for colony having laying workers:

a) Queenlessness.
b) Inability to produce emergency queen cells (no fertilised eggs).

4.10.1.4 Treatment: it is generally agreed that little can be done except to shake the colony out near a strong stock and let them take 'pot luck', always providing that they are disease free.

The following points are pertinent:

a) Difficult (impossible?) to requeen; a colony usually kills an introduced queen.
b) Bees are mostly old and of little use to another colony.
c) If they are united to a queenright colony it has been found that there is the likelihood of them killing the queen of the colony to which they are united.
d) Experiments conducted in France in 1989 on the introduction of queens to colonies of laying workers by dipping the queen in royal jelly and water (70% and 30% respectively) are claimed to be a successful treatment.

4.10.2 The signs (detection) of a drone laying queen and the causes for this failure.

4.10.2.1 The visual signs:

a) Unmistakable worker cells with drone cappings (raised).
b) Presence of a queen (actually seen).
c) Drones produced are small and abnormal (stunted).

4.10.2.2 During the season:

a) Queen produces small areas of drone brood in the middle of large patches of worker brood.
b) As the season progresses, worker brood becomes less and drone brood increases.
c) Because some worker brood remains, it is clear that a queen must be laying.
d) Eventually there will be nothing but drone cappings. At this stage the colony will be reasonably large.
e) Drones are smaller and the abdomen stunted.

4.10.2.3 In the spring:

a) At the first examination of the colony there may be one or two frames of drone brood only (no worker brood present).
b) Is it a drone laying queen or laying workers?
c) If a queen (drone layer), the laying pattern will be orderly i.e. compact patches of brood with very few empty cells.

4.10.2.4 Possible causes for a queen becoming a drone layer:

a) Shortage of sperm - inadequate mating or due to age of queen.

b) Physical inability of queen to fertilise eggs correctly.
c) Genetic fault.

4.10.2.5. Treatment:

a) requeen or
b) unite after removing old drone laying queen.

4.11 The dangers of robbing and how robbing can be avoided.

4.11.1 General points in relation to robbing.

a) In nature, a concentration of colonies does not occur and therefore robbing is not a problem. It only occurs where the beekeeper has concentrated his stocks on to a single site to form an apiary. The beekeeper with only one stock will seldom have trouble with robbing.
b) Robbers are generally bees from another colony but wasps, hornets and ants can also rob a hive.
c) Robbing is for honey only, the other hive products such as pollen and propolis attract no attention as plunder.
d) Different strains of bees have different propensities to rob other colonies, the Italian yellow strains being the worst; they are inveterate robbers.
e) It is more likely to start after a nectar flow has come to an abrupt halt and in times of dearth.
f) It is usually started as a result of bad management practices on the part of the beekeeper.
g) Robbing can occur between hives in a single apiary or between hives in two apiaries.
h) When robbing occurs in an apiary the only method of communication between the bees is by the round dance which only gives information on distance. Because no directional information is available the bees can only search in the near vicinity which may initiate further robbing if a weak colony is discovered.
i) It has been suggested, but not proven, that robbers may release a pheromone to mark the site to be robbed.

4.11.2 Methods to avoid robbing:

a) Prevention is always better than cure, and good apiary practice at all times is usually the answer.
b) Because bees are only interested in a free supply of honey/nectar or sugar syrup available in quantity then there should be no spillage or trace of syrup outside any colony or within the apiary.
c) There should be no way into any colony except via the designed entrance; all equipment should be bee-tight.
d) Colony entrances should be adjusted to the size (or strength) of the colony and to the time of the year and flow conditions.
e) When there is no nectar flow, colonies should not be kept open for too long during manipulations.

4.11.3 Methods of detection of robbing.

There are two types of robbing. The first involves fighting at the entrance of the robbed hive and the second is called silent robbing where no fighting takes place at or within the robbed stock. The behaviour of the foraging bees is quite different in the two cases.

Silent robbing: is characterised by the robbed colony continuing to work normally while at the same time the robbers also enter and leave the robbed colony in a normal manner. The robbed colony can itself be robbing another colony at the same time. The only tell-tale sign is the flight of the bees returning directly to another colony in the same apiary.

Robbing with fighting: has two recognisable characteristics. The first is the fighting outside the robbed hive and the second is the flight of the robber bees approach which is nervous and erratic. The erratic zig-zag flight is curious because it alerts the guards of the robbed colony. Once the robber bee alights and is challenged it becomes submissive and often offers food to the guards.

The characteristic common to both types of robbing is the flight of the laden and unladen bee; rear legs forward in the first instance with a full honey sac and with rear legs trailing astern when unladen in the second instance. The normal rear leg position in flight is reversed, ie. a normal forager should not leave the hive full and return empty.

4.11.4 Methods used to terminate robbing. There is no effective way to stop robbing the day it starts. Removal of the robbed stock to another apiary is unsatisfactory as it usually gets robbed again at the new site (the colony being possibly marked by pheromone). The robbing stock is likely to find another weak stock and continue robbing. The following actions are all effective to some degree:

a) Remove robbers to a remote site isolated from other colonies in the immediate vicinity.
b) Reduce all entrances and make the nucs and weaker stocks a narrow tunnel 2 or 3 inch (50 to 75mm) long.
c) Straw and grass to cover the entrances of both the robbed and robbing hive to confuse both parties has been suggested by some writers.
d) Plain glass leant up against the entrance allowing only entrance from the sides.
e) Reversal of the robber and robbed colony.

If any signs of robbing do occur, we consider that the first action must be reduced entrances and this is why it is so important to have the hive entrance block always stored in the hive diagonally across the crown board when not in use. Nucs are particularly vulnerable and methods of restricting any nuc entrances immediately must be normal apiary management. Note that many of the equipment suppliers, economising on wood, do not make the trim in the roofs of hives deep enough to take an entrance block - they are very easily rectified by tacking 4 laths to the existing woodwork inside the roof.

If a robbed colony is moved it is always wise to leave a frame with some stores in it in an empty hive on the site and allow the robbers to clean it out and finish the robbing job to their satisfaction (the one frame can be put in a spare nuc or travelling box).

One final last point on silent robbing. We discovered, quite by chance, many years ago the robber bees inside the robbed hive passing the spoils to their compatriots through the roof ventilators. Closer observation revealed this to be quite a common occurrence. The solution was simple. We have equipped all our hive roofs with protective wire gauze both inside and outside the hive making contact between the two parties impossible. It also has the advantage of preventing the roof ventilators becoming blocked by other insects.

4.12 One method of uniting colonies.

4.12.1 General considerations. There are various points concerning bee behaviour and beekeeping practice which are of interest before considering the possible methods of actually uniting, these are:

a) Both colonies must be disease free; the spread of disease is caused more often by the beekeeper rather than by any other mechanism.

b)In beekeeping literature mention will be made of 'colony odour' and 'hive odour'. Butler (of queen substance fame) postulated that colony odour is genetically produced and each colony has its own characteristics. On the other hand, Bro. Adam is of the opinion that there is no such thing as colony odour but that there is a hive odour which depends entirely on the materials of the hive and the income (nectar and pollen) which in turn depends on the weather. The hive odour is carried by the individual bees. No one has disputed the concept of hive odour but it has not been subjected to any scientific experiments or proof.

c) During times of dearth there are many guard bees at the entrance of a colony, some of these being potential foragers if forage was available.

d) When there is plenty of forage and a flow on, there will be virtually no guards at the entrance. It is likely that all the colonies in the same apiary are working the same crop and the hive odours are likely to be very similar. Under these conditions, drifting bees are accepted in another colony without challenge or fighting.

From the considerations above it is clear that the best time to unite colonies is during a flow. Feeding, particularly with a scented syrup, when there is a dearth is the alternative solution, although it is not too easy to feed both colonies separately during the uniting process and feeding both separately beforehand is usually the order of the day.

4.12.2 Methods of uniting. There are a variety of ways of uniting, the two more important methods and variations will be described. These are:

a) Newspaper method.
b) Direct uniting.

• Newspaper method. This method is probably the most widely used and is generally very reliable and successful in use. The principle involved is very simple; a queenright (QR) colony and a queenless (QL) colony are joined together with a sheet of newspaper between them. The bees chew the paper away and intermingle slowly and hence unite. The paper is deposited outside the hive in the course of the next 24 hours. Method is described below:

a) The two colonies to be united have to be brought adjacent to one another with their entrances in the same direction. Note that colonies of bees may only be moved 3 feet or 3 miles.

b) The manipulation of uniting colonies should be done in the evening when both colonies have virtually finished flying. The reason is obvious, if the bees are flying then some of the returning foragers will be returning to the entrance of a foreign colony and fighting is likely. Once fighting starts more guards are alerted and then all bees from the other colony trying to enter will be involved. This simple precaution is seldom, if ever, recommended in the literature on practical beekeeping.

c) The newspaper requires 3 or 4 pin holes made in it to help start the process of paper destruction. This can be done with the corner of the hive tool blade if care is used. It is useful to cover the paper with the queen excluder to stop it blowing around during the manipulation. Note the requirement to remove the queen excluder the following day after uniting to release any drones above it.

d) Prior feeding is required if there is no flow on.

e) One colony must be dequeened, the first part of the manipulation. Some books have suggested in the past that the two queens will fight it out and the younger queen will succeed. There is no definite proof that this is so and the possibility exists that the surviving queen may be damaged in the fight. Our advice is do your own selection and be sure of the result.

f) Now comes the last, but vexed, question of which goes on top and which goes below? There is the QR colony and the QL colony and either may be the strong (STR) one or the weak (WK) one. Consultation of 4 books, recommended reading for the BBKA exams, gave the following result:

	BK1	BK2	BK3	BK4
QR or QL on top	QR	QL	QR	--
WK or STR on top	WK	--	STR	WK

The curious thing is that although the authors were recommending a particular approach, not one of them explained why their way was presumably the right way and whether either of the two conditions take preference. If anything preference would be given to the strong colony being above the weak colony on the basis that the weaker would have the minimum guards at the entrance to oppose returning foragers. A case could be made for having the QR colony below with a queen excluder over on the basis that the arrangement can be left for 3 weeks to allow all the brood to hatch in the upper box; the top brood box can then be removed. We are of the opinion that it does not matter which way round they go and any combination will be successful if 3 criteria are observed, namely:

• De-queen one colony.
• Do the manipulation in a flow or with colonies fed for 2 days before the manipulation.
• Do the manipulation just as it is getting dark.

It is not clear where or when the newspaper method originated but it is simple and effective. In some ways it is similar to using a screen between the two colonies for a few days before removing the screen to allow the bees to unite. This method has the disadvantage of requiring a separate entrance for the top colony which is closed when the screen is removed.

• Direct uniting. This is usually undertaken with small colonies (eg. 2 nucs to be united) which may, in total when combined, fill a single brood box. The colonies are brought together, one de-queened and again the operation undertaken in the evening as follows:

87

a) Each frame with bees is removed one at a time from the colonies, dusted with flour or sprayed with a weak syrup and placed in the new brood chamber.

b) The frames are taken alternately one from one colony and then one from the other and placed in the new brood box also alternately so that there is a complete mixing.

c) Care should be taken not to split the brood nests which should combine to make one large one.

d) Finally, the bees are heavily smoked and bumped around to create confusion and the colony closed up.

The flour or the syrup gives the bees an immediate job to do and fighting is non-existent or else minimised.

The words of Bro. Adam, on the subject of uniting, are of considerable interest; exposure to light has a calming effect on bees and when they have been exposed for some minutes, they will peaceably unite without any other precaution throughout the whole season. We follow his wisdom with small colonies and nuclei but use the newspaper method for larger colonies. A further variation on the direct uniting method is to place the frame with the queen and bees in the new brood box and then shake all the other bees from both colonies in front of the hive, placing the empty frames in the brood box. The shaken bees are sprayed with syrup or dusted with flour and allowed to return to the hive. This method is not one that is recommended these days, the job can be done with less confusion and uproar in the apiary.

4.12.3 Other points relating to uniting.

a) Swarms can be thrown together (queens and all) into the same hive within a few days of one another without fighting. When there is a surplus of swarms it is a good way of dealing with them. We have bumped up to 3 swarms all together on the same day and had two brood boxes of foundation drawn out in 2 weeks; when the flow starts such a unit will collect a surplus.

b) Some books state that uniting when there are no drones about is a bad time for this operation; the rationale is not understood. Uniting colonies before winter is a classic time for rationalising the apiary.

c) It is a well known observation that a strong colony will collect more surplus than two weak ones; it is important to ascertain the reason for weakness. If it is disease or poor queens, then uniting will not alleviate the problem.

4.13 The reasons for uniting bees and the precautions to be taken.

4.13.1 The reasons for uniting bees may be summarised simply as follows:

a) Making one strong stock from two weak ones to maximise the honey yield.

b) To reduce the number of colonies at the end of the season going into winter.

4.13.2 The precautions to be taken when uniting have been enunciated in section 4.12 above. We must reiterate the importance of ensuring that the two units to be united are disease free. Samples must be taken from both stocks and examined by a microscopist competent in disease diagnosis.

4.13.3 It has been suggested that a colony of laying workers can be united to a QR colony. Such colonies are virtually impossible to requeen *. We disagree that uniting is a solution because if there are laying workers, the colony will have been QL for 3 weeks or more and all the bees will be old ones. If united satisfactorily they will die off quickly by natural causes and the recipient colony will derive little gain from the addition.

* Recent work in France indicates that it is possible to requeen colonies of laying workers by dipping the queen in a solution of royal jelly (70%) and water (30%) and introducing them directly. The success rate claimed is greater than 70%.

4.14 A method used to clear honeybees from supers.

4.14.1 General points on clearing bees.

a) By definition, clearing implies a crop has been collected and the flow is over; robbing can easily be started unless care is taken when removing the crop.
b) Bees are generally more irritable after the flow and will be more inclined to defend their stores than before the flow finished.
c) Entrances must be reduced at the same time that supers are being cleared.
d) If more than one super has been used it is common for brace comb to have been built joining one super with the adjacent one, the brace comb being filled with honey. It is virtually essential to remove this brace comb 24 hours before clearing to avoid dripping honey from removed and cleared supers. It is a sticky job to do but well worth having the frames cleaned up and no honey dripping while they are being collected. The brace comb should not be there and emphasises the importance of bee space and the many incorrect frames that are in use. The process known as 'cracking the supers' is not given the attention it needs in modern bee literature. It should be done just before dark. It also prevents the surprise of finding the supers not cleared because of brood in the one next to the brood chamber if the first supers are checked during the cracking process.
e) Supers should be removed very early in the morning before the colony has started flying and taken straight to the extracting room for extraction the same day.

4.14.2 Clearer boards. There are basically two types one using Porter bee escapes and the other called the Canadian type with long tunnels for the bees to traverse to get from one side of the board to the other.

• Porter bee escapes: possibly the most popular device in UK for clearing bees. The following are the salient points about its use:

a) The phosphor bronze springs require very delicate adjustment to a gap of $\frac{1}{16}$ inch (1.5mm) and to be free of propolis and wax if they are to work satisfactorily.
b) Two Porter bee escapes per board should be used for rapid clearing and to ensure that if one escape becomes blocked the other one will still be operative.
c) Any clearer board should have an internal bee entrance incorporated in the design with an opening and closing device which can be operated from outside the hive. When wet supers are returned to the hive, the entrance is opened allowing the bees to enter the supers and, conversely,

when it is closed they can be cleared of bees again. It is important that the operating lever allows the roof to be put in place when the supers are off the hive.

d) Approximately 24 to 48 hours are required to clear the supers. The time depends very much on the weather and the flying conditions at the time the board is put on, the better the conditions the shorter the time required to clear.

e) The bee escapes will require cleaning from time to time. Methylated spirit is an ideal solvent for propolis and wax.

• Canadian clearer boards: have the advantage of no moving parts to be propolised and go wrong. The salient points of this mode of clearing is as follows:

a) The same time is required (perhaps marginally shorter) to clear; however, if the weather is bad they are not as effective as the Porter bee escape. The bees seem to learn very quickly that they can return to the supers via the same exit route. The supers must be removed at the latest after 48 hours.

b) An entrance capable of being opened and closed from outside the hive is required identical to the board with Porter bee escapes.

c) If by any chance there may be an odd drone in the supers, they can traverse the exit route without blocking it as would happen with a Porter bee escape.

• 8 way plastic escape: which is pinned to the underside of the board directly below a suitable hole. There are 8 plastic slots for the bees to reach the brood chamber and there are again no moving parts. The principle is the same as the Canadian clearer board. Our findings are that it works no better than the Canadian board.

4.14.3 Shake and brush method. The method appears to be simple and indeed is, if used at the right time.

a) A spare empty super is required, to receive the cleared frames, placed on a roof behind the hive (note the roof is not upturned as most books recommend) with a cover cloth over to prevent any flying bees re-entering the cleared frames. The colony is smoked first at the entrance and then at the top to drive the bees downwards in the supers. One frame at a time is shaken free of bees and those remaining on the frame are brushed off with a feather. The frame free of bees is then placed in the empty super. As one super is cleared so that becomes the receptacle for the next and so on.

b) If the supers are sealed smoking has little effect on the bees; they are only subdued when they are able to gorge themselves with honey, these bees, unable to gorge themselves are not immediately subdued, only driven downwards. After the honey flow has ceased the colony is likely to be more aggressive and will defend their stores. It will be clear that it is not a method to be used by the uninitiated at the wrong time and certainly not in an urban situation.

c) Where should all these bees be shaken? We like to shake them back into the hive rather than at the front, only the bees which we brush off land up at the front. The reason for this is that we keep the super covered with a cover cloth except when a frame is being shaken and if the bees are in the hive, we are in control of the situation and not the bees.

d) The final consideration is when should this method be used? At a time when the bees are not flying to safeguard against robbing being started; this means early morning or late evening.

4.14.4 Other clearing methods. To provide an overall picture other methods of clearing should be noted, these are:

a) Mechanical blowers usually powered by electric motor which in turn is powered by a portable generator. Not for the small time beekeeper.

b) Chemical repellents. The three most commonly known are:

- Carbolic acid - not used these days.
- Butric anhydride - popular in USA.
- Benzaldehyde (smells of oil of bitter almonds) - used quite extensively in UK and works well. Should be kept in the dark.

4.15 The process of extracting honey from combs and a method of filtration and bottling of honey suitable for a small scale beekeeper.

Extracting and bottling of honey on a small scale involves decapping super combs (4.15.1), types of extractor and extraction (4.15.2), straining and storage (4.15.3) and finally the bottling process (4.15.4). Each is examined in the paragraphs below.

4.15.1 Suitable methods of decapping super combs.

4.15.1.1 General. There are a variety of methods for decapping combs, some for the hobbyist and others for large scale operation where a permanent installation is necessary. Some of the more common methods are as follows:

a) Various types of knives used either hot or cold and with sharpened edges or serrated cutting edges. The common feature is the length of blade which requires to be about double the width of the super frame.

b) Heating for the knives can range from the simple method of dipping into hot water to electric elements designed into the blade or steam being passed through the blade.

c) The electric carving knife with the two reciprocating blades adjacent to each other works well if cleaned after each cut in hot water. This is our particular choice.

d) A small gas blow lamp or hot air paint stripper works well and obviates having to process the cappings which melt around the edges of the cells. The idea came originally from Norway. We have used it successfully on various occasions.

e) Decapping machines based on a heated reciprocating blade in a fixed position, the comb being passed over the blade. For larger scale operations.

f) Flailing machines which rip the capping off are also very efficient for large scale operations.

g) There are a variety of multi-pronged forks for scraping off the cappings, useful on badly drawn comb.

All the methods require some form of tray or receptacle to receive the cappings and process them.

4.15.1.2 Uncapping trays. The essential feature of these trays is to provide a receptacle for the cappings to fall into in order to separate the honey from the wax capping. The two basic methods are cold straining through a suitable gauze or melting the cappings with the honey and allowing them to separate on cooling.

a) The Pratley (or Prattley; there is doubt about the spelling) tray uses the heating principle. It is constructed in stainless steel with a water bath and heating element on the underside. It is expensive and probably one of the worst designed pieces of equipment available. There is no thermostatic control for the heater so one is continually switching on and off. The tray is built on a slope with the thin part of the wedge, and minimum thermal mass, at the place where maximum heat is required. With the result the whole thing clogs up and slows down the extracting process. The separated honey is heated to such an extent it is only good for cooking.

b) The cold straining method seems to be as good as any, the honey is not ruined and the cappings can be washed for mead making or given back to the bees to clean up.

c) None of the devices makes provision for locating the frame over the uncapping tray and it is necessary to fix a bar across with a suitable circular hole about $^1/_4$ inch (6mm) deep to rest the lug of the frame in while actually cutting the cappings off.

d) The beginner to beekeeping should be warned of the pitfalls in this area before he parts with his money; unfortunately most of the experienced beekeepers have learnt the hard way.

4.15.1.3 The actual operation of removing the cappings.

a) The important thing is to arrange the set up so that the actual operation is undertaken comfortably with no strain, with everything at the right height and everything to hand. It is as well to study the flow of the work from the full supers coming into the extracting room to the empty ones going out.

b) Having two people on the extracting makes life much easier, one decapping and the other operating the extracting machine.

c) Most of the books recommend that the frame is held at an angle of about 30° to the vertical and the cut made downwards away from the end of the frame which is being held. The other way is to cut upwards (which we find easier). In both cases the cappings fall away from the frame into the uncapping tray and do not stick to the uncapped comb.

d) Again, most of the literature states that the cut should be made just under the capping (in the air space) to minimise the amount of honey removed from the comb. This also is considered to be a matter of preference; we do a straight and level cut across the face of the comb to end up with a set of even combs in the super ready for next year's operations. There are always likely to be some uneven combs and this is a good time to put these to rights.

Manley type frames help to provide a straight cut.

e) If the knife is heated in hot water it is as well to have a clean damp cloth close by to wipe the knife dry before making a cut, continual drops of water will increase the water content of the honey.

4.15.1.4 Other points about decapping.

a) Wax cappings should be separated from the honey and rendered down carefully and the wax used for showing. It is unadulterated with other hive products and requires little cleaning.

b) The cappings can be cleaned up by the bees in say a Miller type feeder with a grill to allow the bees to access both sides. They may require a stir after a couple of days to let the bees get to the inner sticky portions.

c) Extracting is a sticky thankless task and care should be taken to keep everything scrupulously clean and tidy.

d) Any honey which has been heated in the Pratley tray should only be used or sold as cooking honey (baker's honey).

4.15.2 The principles of honey extractors, both tangential and radial.

4.15.2.1 General considerations. The principle involves two forces which occur when an object is rotated eg. a ball on a string. The first is the centrifugal force which acts away from the centre of rotation and the second is the centripetal force which acts towards the centre of rotation. In the case of the ball, the centrifugal and the centripetal force act on the ball and because they are equal the ball stays where it is on the end of the string. In the case of a honey extractor, only the centrifugal force acts on the honey and it moves outwards away from the centre, hits the drum wall and trickles down to the bottom under another force, gravity. The centripetal force, of course, is acting in the rotating frame of the extractor and being rigid does not move.

4.15.2.2 Tangential extractors: generally have a cage whereby the frames for extracting can be inserted at a tangent to the circle of motion allowing only the honey on the outside face to be removed during operation of the extractor. The honey on the inner face is pushed on to the comb septum during operation thus necessitating the frame to be reversed in order to complete the extraction. The salient points of this type are as follows:

a) With full frames the first side can only be partially extracted at slow speed otherwise the weight of honey on the inner face is likely to break the comb.
b) The combs must be reversed to extract the second side and then reversed again to complete the extraction of the first side. This is time consuming.
c) It is virtually essential to have wired combs for use with this type of extractor.
d) The number of frames that can be extracted is limited by the size of the cage, 6 being the realistic limit and 3 or 4 the norm.

4.15.2.3 Radial extractors: are all very similar with a rotating framework designed to hold the frames radially with their top bars vertical and parallel to the sides of the drum and their side bars on a radius of the extractor's rotating framework. The salient points are:

a) Honey on both sides of the comb is extracted simultaneously and there is no reversing of frames which makes the process that much quicker.
b) Combs that are unwired can be easily extracted without damage.
c) The design permits a greater number of frames for a given diameter drum compared with a tangential, 10 being a realistic size for a small model (or one super).
d) Speed control is not as critical as the tangential but nevertheless it is necessary to start slowly.
e) For a given speed and number of rotations, the radial is inferior due to the amount of honey remaining in the comb.

4.15.2.4 Other points of interest relating to extractors.

a) Materials for construction range from tin plate, galvanised steel and polythene to stainless steel. These days only the two latter materials are used. Stainless steel costs more but is likely to prove the most economical in the long run.
b) Both types can be either hand or motor driven. If the electric motor is used, a speed control device is necessary to give a range of speed from 0 to about 400 rpm. The problem arises in getting a device to give high torque at low speed when it is starting up from rest. The horse power (hp) required from the motor is quite small, a fraction of a hp being adequate.
c) When purchasing a radial type, care should be taken that it will accommodate the frame the beekeeper intends to use. Many are designed for $^7/_8$ inch (22mm) side bars and the full complement of Manley frames cannot be loaded because of their wide side bars.
d) Most tangential types can take a reduced number of brood frames which is often useful; only some of the radial types can do this.
e) Both types require to be loaded giving the best dynamic balance. No matter how carefully the frames are selected and loaded it will never be perfect and the extractor at top speed wants to move. There are two solutions, one to screw it down to a platform with castors and let it move or two, to bolt it securely to the floor. Our own is fixed in three places with rigging screws and chain from the top rim of the extractor.
f) If a fixed installation is preferred then it is desirable to raise it so that a honey bucket can be put straight under the outlet tap without any lifting.

4.15.3 Methods of straining small quantities of honey and its subsequent storage.

4.15.3.1 The objective: of straining is to remove solid matter down to a particle size determined by the strainer. There are three types of solids, those that sink to the bottom, those that float and those that remain in suspension. The solids can include wax, bees, grubs, propolis, sugar crystals, wood chips and other extraneous matter. Different strainers will remove different solids.

4.15.3.2 Types of strainer include the following methods:

a) Single settling tank. The tap is at the bottom and only those solids that float will be strained out including air bubbles.

b) Sump tank with baffles. The input is usually at a low level and the outlet at a high level of a tank containing 3 or 4 baffles giving 4 or 5 compartments. The baffle openings are alternately top and bottom with the honey flowing alternately under and then over the baffles. The surface can be skimmed as required to remove floating debris while the dense solids collect in the bottom of the tank. The tank can be double walled to provide a water jacket for heating if required. Essentially for the larger scale producer.

c) Wire and cloth strainers. These can be in a variety of formats depending on the scale of the operation. Wire strainers should be made of Monel metal or stainless steel and cloth strainers are made of scrim, silk or nylon. Commercial strainers eg. O.A.C. honey strainer, incorporate a series of concentric strainers of different mesh size the honey flowing through the coarse mesh first and the finest mesh last eg. 12 mesh / inch to about 80 mesh / inch. All straining should be below the surface of the honey to prevent bubbles forming.

d) The most simple type used by many small time beekeepers is to strain the honey as it goes into the settling tank. It has one major drawback; it produces a large number of bubbles which are formed when the honey drops from the straining cloth into the settling tank. The air bubbles can be reduced by using a long conical shaped bag strainer.

4.15.3.3 Other points in relation to straining are:

a)The higher the temperature the easier it is to strain with a fine mesh. A temperature of 95° to 100°F (35° to 38°C) is considered to be satisfactory.

b) If a honey such as rape is to be stored in buckets and bottled later it is better to complete the fine straining before it is stored. Later it only needs warming to a point where it will flow for bottling; this stage is reached well before all the crystals have melted. In this state it would be impossible to pass it through a strainer without bringing it back to the completely liquid state and thereby heating it unnecessarily.

c) Long straining cloths can be an advantage; when clogged, pull across the tank to an unused portion.

d) Large commercial honey packing organisations use very fine filters by pumping the honey at high temperatures and then cooling it quickly after straining. Such methods are not a practical proposition for small scale operations but the liquid honey has a wine like clarity which is very attractive to the buyer.

4.15.3.4 Subsequent storage of strained honey. Honey in small quantities should be stored in 30lb polythene (white only for food) buckets with a well fitting air-tight lid, anything bigger is difficult to melt down again for further treatment and bottling. The old method used 28lb tins with lever lids with attendant problems of the lacquer and plating becoming faulty and the subsequent rusting.

The main points are:

a) The buckets require to be as full as possible to minimise the amount of air trapped in the top.

b) Before the lid is snapped shut, the centre of the lid should be depressed onto the honey again to minimise the air content.

c) Store at a temperature of 57°F (14°C) for rapid granulation and then at as low a temperature as possible after it has set. Don't open to check the granulation, do it by feel, the sides of the buckets are quite flexible when the honey is in the run state but very solid when granulated.

d) An old refrigerator makes an ideal warming cabinet for reheating when fitted with a fish tank thermostat, a small fan and a couple of electric light bulbs. The fan is necessary to prevent hot and cold spots forming. Ours used to double up as an incubator for queen cells in the summer by resetting the thermostat before we found an old steriliser which has been converted into an incubator.

4.15.4 Methods of small scale bottling and preparation of honey for sale.

4.15.4.1 The types of honey: which are likely to be bottled on a small scale are as follows:

a) Run or liquid honey. Because all honey is a supersaturated solution of sugars it will granulate in a greater or lesser time depending on the types of sugars it contains. It is therefore essential that it is heat treated before sale to delay granulation for 3 to 6 months (130°F or 54°C for about 45 minutes for an average honey). Partially granulated run honey looks terrible and is a very common fault with the hobbyist beekeeper.

b) Granulated honey. This should be of a fine texture and an average honey should be seeded with 10% rape or similar honey which granulates with a fine grain.

c) Creamed honey. Again a fine texture is necessary.

d) Chunk honey. The comb must be surrounded with a heat-treated liquid honey.

4.15.4.2 Faults that are to be avoided. There is much local honey on sale which should really be taken off the shelf and returned to the beekeeper as being unfit for sale. Many of these jars are labelled with a county association label and thereby brings other beekeepers a bad reputation. We believe that a county organisation should take some responsibility for quality if their label is to be used. The common faults are:

a) Partial granulation of run honey.

b) Frosting of granulated honey. This a particularly difficult fault to avoid with honey that granulates rapidly - it is caused by shrinkage, generally at the neck of the jar, as the honey granulates and is associated with low temperatures. Honey that granulates slowly or granulates at higher temperatures (greater than 57°F or 14°C) seldom exhibits this fault.

c) Fermentation. A vertical streak in granulated honey with the surface texture rough. In run honey recent bubbles on the surface can be seen. At an advanced stage it has a characteristic unpleasant smell.

d) Scratched lids and unclean wads.

e) Underweight.

f) Incorrect labelling.

4.15.4.3 Methods to be employed in order to produce a quality product:

The first requirement is that all honey must be ripe when extracted otherwise fermentation can occur. Heating the honey to a temperature of 130°F (54°C) for a short time will kill the yeasts but will not reduce the water content sufficiently to make it completely safe from re-infection by other wild yeasts.

Run or liquid honey:

a) Usually bottled and heat treated after it is in the jar by putting the jars complete with wads and lids in a water bath for about 45 minutes at 130°F (54°C). The same effect can be obtained by putting the jars in a microwave oven covered with 'clingfilm' in lieu of the metal lids. It will be necessary to experiment to get the power and time right to obtain the correct temperature.

b) Clean jars are essential, any minute particles of dust or other solids provide a nucleus for granulation to commence.

c) A few jars should be filled after taking the weight of the jars and then re-weighed. Such a test should be carried out at random during bottling. If no light can be seen through the jar under the lid when it has been filled, it will almost certainly pass the weight test.

Set or granulated honey:

a) Granulated honey should have a fine texture; coarse crystals give the honey a gritty characteristic which is generally not liked by the consumer. Therefore most main crop honey should be seeded with a rape or clover to obtain the fine grain.

b) The seed should be melted to a point where it just flows and has a granular texture. 10% should be stirred into the liquid honey (which has been warmed to the same temperature as the 'seed') without causing bubbles, and then bottled as for run honey.

c) To ensure granulation takes place as quickly as possible the jars should be stored at a temperature of 57°F (14°C) or as near as possible. Fine granulation minimises the water content between the crystals and thereby minimises fermentation.

Creamed honey:

a) Is produced in exactly the same way as finely granulated honey except that after it has fully granulated and set hard a further treatment is given.

b) If the set honey is heated to 80° to 90°F (27° to 32°C) for approx. 24 hours it will become creamy and smooth. It is then stirred with a special mixing tool avoiding the introduction of air bubbles and then allowed to re-set. After this treatment it will not revert to the hard state.

Chunk honey:

a) Requires one piece of sealed comb approx. 1 inch×1 inch (25mm×25mm)× the height of the jar and filled with heat-treated liquid honey.

b) The major fault is to provide too small a piece of comb. The liquid honey needs to be well filtered and 'bright' to show off the chunk in the jar.

Equipment required:

a) Warming cabinet. 30lb buckets take about 5 days at 90°F (32°C) to get to a state where the contents will flow eg. rape. For other honey, to get it to the stage where it is completely liquified ready for straining and bottling about 3 days at 120°F (49°C) is required.

b) A good set of scales is required and the calibration should be checked regularly.

c) Settling tank to hold about 112lb, together with honey stirrer for stirring seeded honey.

4.15.4.4 Other points on bottling.

a) When bottling has been completed the jars should be free from stickiness on the outsides.

b) Suitable labels should be attached on each jar; note many labels are too big for $^1/_2$lb jars.

4.16 The need for good hygiene in the handling of honey for human consumption.

The need for good hygiene in the handling of honey for human consumption is really common sense and can be defined under the two following headings, namely:

1. To ensure that the beekeeper does not infect the final packaged product with foreign bodies of any kind, particularly those organisms appertaining to diseases of mankind.

2. To ensure that the beekeeper does not run foul of the law and be subsequently prosecuted under United Kingdom or European Economic Community (now European Union) legislation which is generally administered and policed by environmental health officers and personnel.

Both the above are inter-related by virtue of the fact that the legislation is drafted to avoid problems of foreign bodies in a natural food.

Most beekeepers undertake the processing in the family kitchen. It prompts the question whether it is suitable? Are there two sinks (one for washing utensils and the other for washing hands) and are surfaces of the walls and floors washable? It is likely to be taboo if there is a washing machine installed where soiled linen is washed despite the fact that all the family food is prepared every day in the same room. Beekeepers must be vigilant of current legislation, it is all too easy to transgress unwittingly. In the future EEC regulations may require all beekeepers producing, processing and selling honey to be registered.

4.16.1 The common sense approach.

Tools, utensils, equipment and packaging. This is a straightforward washing up job with lots of hot water and clean cloths. No soap or detergent should be used to ensure that no contamination can occur with cleansing agents of any kind. The packaging of the honey in glass jars and plastic cartons is the real danger area as most beekeepers assume that they are clean when received from the supplier. In many cases they are bought in bulk from the manufacturers and boxed up into smaller quantities by the bee appliance suppliers. The boxing up process often introduces foreign bodies and dirt. Beware of small chips of glass in honey jars and cracked jars; they should all be turned over to remove any chips and then washed and polished before filling. Even when the jars arrive sealed in polythene the same care should be given to the jars before use. Lids will also require inspection. A good deal of work can be eliminated if the jars and lids are stored in a clean dry area before use. The extractor, filters, settling tanks and storage containers should be cleaned of all honey immediately after use, dried and stored carefully away until required.

Hands and clothing. It is understood that most people catch a cold or flu or other viral infection by infecting themselves with their hands which in turn have picked up the infection from another source. Consider the common cold; it would be an unusual individual who did not infect their hands by using a handkerchief or tissue. These infected hands then use the handle of a door covering it in large quantities of 'bugs'. Along comes the next person to use the door handle and pick up the 'bugs' on their hands. They infect themselves by touching their face, mouth, eyes, etc. Shaking hands with somebody who has a cold is a fine way of catching it from them. A grand old merry-go-round. Wash your hands and then don't touch any part of your face. If you do, then wash your hands again before continuing with the task of processing honey. Use a clean overall especially for the job; regulations these days would require all operators, male or female, to wear a suitable protective hat. Smoking is, of course, a no go activity.

Clean area. The work area, which for many hobbyist beekeepers will be the home kitchen; very few will have a discrete extracting and honey processing room custom built for the purpose. Do ensure that everything is spotless before operations start paying particular attention to all working surfaces. All pets should be banished. Windows and doors, if not screened, should be kept closed during extraction and bottling of honey. There should be no trace of vermin, flies, bees, wasps, ants, spiders etc. either in the extracting room or in the room where the honey is stored. No other activity should be taking place in the room where the honey is being processed.

Personal hygiene. Any one selling honey to the public may have their premises inspected by the environmental health inspector. Personal hygiene is paramount, dishevelled hair, dirty unkempt nails and hands (broken skin), stained clothing, signs of smoking, obnoxious body odours or signs of ill health eg. coughing and sneezing, should not be tolerated by anyone assisting or operating the extracting and bottling process. There should be a separate toilet close by the extracting or bottling room with hot and cold running water where the operators hands can be washed after each visit.

4.16.2 It is very important to keep up to date with the regulations which are constantly changing to 'harmonise' (a popular bit of Euro-talk) them throughout the Community. The BBKA publicise these changes in BBKA News and attention is brought to the County Association Secretaries and to Branch Secretaries. If you are in any doubt, ask your Branch Secretary.

4.17 The legal requirements for the labelling and sale of honey.

The following is only a summary of the requirements for labelling and the sale of honey and is by no means exhaustive.

4.17.1 The following Acts and Regulations apply; they should be referred to if an exact wording is required:

- Food and Drugs Act 1955
- The Labelling of Food Regulations (1970 as amended)
- The Honey Regulations 1976
- The Materials and Articles in Contact with Food Regulations 1978

- The Weights and Measures (Marking of Goods and Abbreviations of Units Regulations 1975 as amended). A European directive came into force on 1st October 1995 replacing the imperial weights by the metric system.
- The Weights and Measures Act 1963
- Trade Descriptions Acts 1968 & 1972
- The Trade Descriptions (Indication of Origin) (Exemption No. 1) Directions 1972
- Consumer Safety Act 1978 Glazed Ceramic Ware (Safety) Regulations 1975
- The Food (Lot marking) Regulations 1992
- The Food Safety Act 1990
- Food Premises (Registration) Regulations 1991
- Honey Regulations 2003*

There are numerous regulations dating back to 1955 made to protect the public from unscrupulous vendors of honey who may sell short-weighted jars with misleading labels containing contaminated honey. From time to time a case is tried in the courts of law. The beekeeper, if found guilty, has been fined and his good name is lost.

4.17.2 Legal definitions.

- "Honey" means the fluid, viscous or crystallised food which is produced by honeybees from the nectar of blossoms, or from secretions of, or found on, living parts of plants other than blossoms, which honeybees collect, transform, complete with substances of their own and store and leave to mature in honeycombs.
- "Comb Honey" means honey stored by honeybees in the cells of freshly built broodless combs and intended to be sold in sealed whole combs or in parts of such combs.
- "Chunk Honey" means honey which contains at least one piece of comb honey.
- "Blossom Honey" means honey produced wholly or mainly from the nectar of blossoms.
- "Honeydew honey" means honey, the colour of which is light brown, greenish brown, black or any intermediate colour, produced wholly or mainly from secretions of or found on living parts of plants other than blossoms.
- "Drained honey" means honey obtained by draining uncapped broodless honeycombs.
- "Extracted honey" means honey obtained by centrifuging uncapped broodless honeycombs.
- "Pressed honey" means honey obtained by pressing broodless honeycombs with or without the application of moderate heat.

4.17.3 Methods of sale.

When sold by retail not prepacked eg. from a bulk container, honey should be sold by net weight. When prepacked ready for retail sale in a quantity of more then 28g, the net weight of honey in the container should be one of the following: 57g, 113g, 227g, 340g, 454g, 680g or a multiple of 454g. Chunk honey and comb honey may be packed in any quantity.

4.17.4 Marking on containers.

Honey should be prepacked for retail sale, or otherwise made up in a container for sale, only if the container is marked with the following information:

* For details see insert in *Honey Marketing* by H. Riches (BBNO)

a) An indication of quantity by net weight in metric units, this may be followed by imperial units.

b) The name or trade name and address of the producer, packer or seller.

c) A description of the honey in one of the following forms:

i) Honey   ii) Comb Honey   iii) Chunk Honey   iv) Baker's Honey or Industrial Honey.

v) The word "honey" with a regional, topographical or territorial reference eg. Devon Honey, Honey from South Devon, Moorland Honey.

vi) The word "honey" with a reference to the blossom or plant origin eg. Heather Honey, Lime Honey.

vii) The word "honey" with any other true description, eg. Honeydew, Pressed Honey, Set Honey.

In wholesale transactions of containers of a net weight of 10kg, or more, a separate document showing the required information is sufficient if it accompanies the container.

4.17.5 Methods of marking containers of honey:

• The metric indications of quantity and imperial units should be of equal size. The metric unit should be placed before the imperial unit. The minimum height of any figure used is:

For a quantity of 28g - 2 millimetres.
For a quantity of 57g or 113g - 3 millimetres.
For a quantity of 227g, 340g, 454g, 680g or 908g - 4 millimetres.

The units of weight used should be at least half these heights.

• The two quantity indications ie. metric and imperial, should be distinct but in close proximity, the metric quantity being shown first. Nothing should be inserted between them.

• The permissible units of weight with their permitted "abbreviations" are:
kilogramme - kg; gramme - g; pound - lb; ounce - oz;
No other abbreviation should be used.

• All required markings should be clear, legible, conspicuous and indelible.

4.17.6 Containers.

Containers should be made of materials, which under normal and foreseeable conditions of use, do not transfer their constituents to the honey in quantities which could endanger human health or bring about a deterioration in its aroma, taste, texture or colour or bring about an acceptable change in its nature, substance or quality. This applies to containers which are in contact with the honey and to the containers which are likely at some later time to be in contact with the honey.

Certain ceramic materials may present particular risks. Packers are asked to obtain an assurance from their suppliers that containers comply with 'The Materials and Articles in Contact with Food Regulations 1978' and 'The Glazed Ceramic Ware (Safety) Regulations 1978' and 'The Glazed Ceramic Ware (Safety) Regulations 1975', if applicable.

4.17.7 Misdescription.

There are two basic types of illegal misdescription; the direct and the indirect or misleading. The direct misdescription should be obvious and can be fraudulent. A simple example would be to describe Australian honey as "Devon honey". Careful thought will avoid indirect misdescription. Examples of such misdescription could be:

> a) An illustration of bees collecting nectar in a moorland setting on honey which is not from moorland.
> b) The statement "Produced in Devon" applied to honey which is blended in Devon from honeys of various origins which may or may not include Devon.

The following guideline should be followed.

> • Any reference, direct or indirect, in words or by means of any pictorial device to the blossom or plant origin should only be applied to honey derived wholly or mainly from the blossom or plant indicated.
> • Any such reference to the regional, topographical or territorial origin of the honey should only be applied to honey which originated wholly in the region, place or territory indicated.
> • Description which have legal definitions should be applied only to products which fall within the generally accepted meanings of those description references.
> • Descriptions and other references which have no legal definitions should be applied only to products which fall within the accepted meanings of those descriptions or references.
> • A honey may fall within more than one definition, in which case it may be described as being any one or more. For example, a pressed Devon heather honey may be described as "Honey" or "Heather honey" or "Pressed heather honey" or "Devon blossom honey" or any other true combination of words.
> • Honey produced outside the UK which has a UK name or mark should be accompanied by a conspicuous indication of the country in which the honey was produced. Blends of honeys from two or more countries, which may include the UK, may be accompanied instead by a conspicuous indication that it was produced in more than one country.

4.17.8 Composition of honey.

> • There should be no addition of substances other than honey.

> • The honey should as far as practicable, be free from mould, insect debris, brood and any other organic or inorganic substance foreign to the composition of honey. Honey with these defects should not be used as an ingredient of any other food.

> • The acidity should not be artificially changed. There is a legal maximum level of acidity ie. not more than 40 milli-equivalents acid per kilogram.

> • Any honeydew honey or blend of any honeydew honey with blossom honey should have an apparent reducing sugar (invert sugar) content of not less than 60% and an apparent sucrose content of not more than 10%. Other honeys should have an apparent reducing sugar content of

not less than 65% and an apparent sucrose content of not more than 5%.

• Honey with a moisture content of more than 25% should not be supplied.

• The maximum water insoluble solid content is:

for pressed honey: 0.5%
for other honey: 0.1%

•The maximum ash content is:

for honeydew honey and blends containing honeydew honey: 1.0%
for other honey: 0.6%

4.17.9 Baker's or Industrial honey.

Honeys of the following descriptions should be labelled or documented only as "baker's honey" or "industrial honey":

• Heather honey or clover honey with a moisture content of more than 23%.
• Other honey with a moisture content of more than 21%.
• Honey with any foreign taste or odour.
• Honey which has begun to ferment or effervesce.
• Honey which has been heated to such an extent that its natural enzymes have been destroyed or made inactive.
• Honey with a diastase activity of less than 4, or, if it has a naturally low enzyme content, less than 3.
• Honey with an hydroxymethylfurfuraldehyde (HMF) content of more than 80mg/ kg.

IMPORTANT NOTE.

Government regulations are constantly changing and being updated. It is therefore important that beekeepers make themselves familiar with up to date information before processing, packing and selling honey for retail sale. Since 1992 many changes have been made as a result of the EEC. Further changes are extremely likely to occur .

4.18 An elementary account of the harvesting of beeswax.

4.18.1 Types of wax collected: during the season are cappings during extraction, brace and burr comb collected during inspection and manipulations and finally old combs which are being discarded.

• The cappings are of high quality and need very little treatment to clean them up for use. They should be used solely for show wax, making cosmetics eg. cold cream and for high quality wax blocks for sale.

• Quite a considerable amount of brace and burr comb is collected during colony manipulations and should be lumped in with the old combs for rendering. Since these two sources contain wax contaminated with propolis (which cannot be removed) it is only suitable after home rendering and cleaning for making foundation, candles, etc. There is no way of cleaning wax at home comparable with large scale commercial operations with heated pumps and filters.

4.18.2 Processing cappings.

• The cappings will be initially separated in a decapping tray or similar device with a mesh basket to allow the honey to drain off.

• The cappings can be given back to the bees to clean up or washed to make mead with the washings.

• Finally the cappings are melted and filtered through lint as a final cleaning process. The wax should not be heated above 150°F (65°C) to prevent discolouration. To save the natural colour of the wax, metal containers should be avoided.

4.18.3 Processing old comb.

• After the wax has melted it adheres to the old larval skins and, except by pressing or centrifuging at high temperatures, some wax will inevitably be lost. These two methods are generally unsuitable for home operation.
• There are two suitable methods for home use, these are the solar wax extractor and the steam boiler.
• The steam wax extractor is a boiler with a mesh cage suspended inside with a drain at the bottom for the wax to run off. The device has a water reservoir which is converted to steam, which melts the comb and wax inside. It can be driven by gas or electricity.
• The solar wax extractor is probably the best device; it costs nothing to run and can be made for a few £s. They are expensive to buy from the equipment suppliers. Most of the double glazing merchants have lots of old panels available which they are only too glad to sell and produce a better job than the commercial models available.
• A second melting in a saucepan of soft water leaving it to cool and float on the water. Any dross ('slumgum') can be scraped off the bottom and it is ready for use. Further cleaning using a filter may be necessary.

4.18.4 The solar wax extractor. As this method is by far the most suitable and there is little meaningful information in the general literature, a few points of interest on the device are:

• When it is normal (at right angles) to the sun on a cloudless day the energy collected is approx. 1kW per metre square (compare with a 1kW electric fire).
• The correct angle to the horizontal (α) is given by the formula: α = Lat - Dec where Lat = latitude and Dec = sun's declination. The sun's declination varies from 0° in March to +23° in June to 0° again in September. It goes on to -23° in December (the winter solstice) and then to 0° again the following March. The average value for the summer months March to September is c. 14°. On the South Coast of UK with a latitude of say 50°, then α = 36°.
• Black bodies absorb the most radiation, so the inside should be black or as dark as possible, and

not white, for the greatest efficiency.

• The glazing should be double and preferably a sealed unit and very good insulation (fibreglass for lofts) is also necessary for efficiency.

• Take care not to burn yourself on the metal inside, the temperature is well over the boiling point of water on a good day.

• Because the temperatures are so high it acts as a steriliser and will kill most pathogens. It is therefore a good idea to make the extractor large enough to take one or two whole brood frames complete with comb. To ensure that foundation wax is free from AFB and EFB pathogens, the wax should be maintained at 100°C for 30 to 60 minutes and this should be done as a separate operation outside the solar wax extractor if there is any doubt, particularly if colonies have been treated with antibiotics for EFB.

• Keep the glass clean for maximum efficiency. The inner surface is the most difficult to clean as it develops a thin film of vaporised wax and propolis; methylated spirit will clean it off.

4.18.5 Other points of interest.

• One very old type of extractor (the M & G extractor) is still to be found and works well with care and controlled heating. It is a metal drum and old comb is put inside and filled up with water and a filter is tied across the open top. Around the outside, and as part of the device, there is a large rim to catch the contents as they are pushed through the filter. The rim has a spouted outlet for the wax. The wax is forced out by hydraulic pressure when water is poured into the high spout, the bottom end of which is connected into the lower part of the drum. If care is not used it is claimed that the ceiling is likely to receive a wax treatment.

• Physical properties of beeswax are:
    - Specific gravity = 0.95.
    - Melting point = 147.9 ± 1°F (64.4 ± 0.55°C).
    - Solidifying point = 146.3 ± 0.9°F (63.5 ± 0.5°C).
    - Insoluble in water.
    - Soluble in chloroform, ether, benzene, etc.
    - When stored cold for some time it develops a surface bloom which is not a mould or
    mildew, the cause of this film on the surface of the wax is not yet understood.

• Most appliance suppliers will purchase rendered beeswax or exchange it for equipment.

** ** ** **

5.0 DISEASE, POISONING AND PESTS

5.1 The appearance of healthy brood and how it differs from diseased brood or chilled brood.

Brood means all stages of the brood from eggs through larval stages to sealed pupae; simply all open brood and sealed brood. We are concerned here with the determination of healthy, diseased or chilled brood by normal eyesight and not by using optical aids such as magnifying glasses and microscopes. It is not possible to differentiate between a normal egg and one that may be faulty, therefore, the following discussion is concerned with larvae of all ages and sealed brood before emergence.

5.1.1 The appearance of healthy brood.

- Larvae: All larvae, both worker and drone, from the newly hatched egg to the fully grown larvae, just before cell sealing takes place, are pearly white in colour, shiny and lie in the bottom of their cells in a curled up position. It is only possible to see the upper side of the larvae, the lower side is floating in the brood food at the bottom of the cell.
- Sealed brood: Has convex coffee-coloured cappings made from a mixture of beeswax and pollen ie. porous to allow respiration to proceed normally. Because the cappings contain pollen they are of a dull matt appearance with no trace of shininess.

Diseased and chilled brood will be directly compared with the norm as outlined above. It is extremely important that all beekeepers are capable of recognising healthy brood and thereby being able to quickly recognise disease problems while undertaking normal colony manipulations.

5.1.2 The appearance (signs) of diseased brood.

- Larvae: All diseased larvae eg. stone brood and EFB are coloured and distorted, unlike the shiny, white and circular appearance of healthy larvae positioned in their cells. There are three exceptions. The first is AFB where the larvae before sealing appear to be perfectly normal; the changes happen when the larvae die after the cell is sealed. Similarly in the case of sac brood the sealed cell is uncapped to reveal a larva flattened in shape with an upturned head (known as the Chinese slipper effect). In the case of chalk brood the unsealed cell reveals a mummified larva.
- Sealed brood: The appearance of the cappings associated with diseased brood is concerned with AFB, chalk brood and sac brood. Any cappings which do not conform to those of healthy brood will have some or all of the following characteristics: sunken or concave cappings, perforated cappings, discoloured cappings and cappings with a darkened damp appearance. Often the cappings are removed entirely revealing the dead larvae or pupae before the bees have had time to remove them. It is instructive to uncap sealed brood around the cell of a dead larva eg. chalk brood, as it often reveals other larvae dead in their cells before the bees have uncapped them. The bees are able to determine whenever there is a dead larva in a sealed cell.

5.1.3 The appearance (signs) of chilled brood.

Chilled Brood is brood in all stages which is killed due to exposure to low temperatures. All stages means from the hatched larva to the sealed pupa and for this reason it is very easy to diagnose as no other diseases kill brood of all stages in one fell swoop. It will be clear that it is not a disease but a condition, as no pathogen is involved. Dr.Bailey states that unsealed larvae can survive several days at room temperature of c. 65°F (18°C), so the temperature drop must be quite severe or prolonged to kill them in a colony. We have never seen chilled brood in a colony but have produced it artificially in the refrigerator.

5.1.3.1 Causes.

- When a colony is approaching starvation (there is no carbohydrate to convert into heat energy).
- Due to spray poisoning (many bees lost).
- Stated in many books to be due to mishandling by the beekeeper (opening a colony for too long in low temperatures).

5.1.3.2 Signs.

- Brood of all stages dead.
- Dead brood at the periphery of the brood nest.
- Some of the capped cells may be perforated.
- Larvae turn grey and then to black in colour and remain shiny though they are discoloured. This is a very important sign and is discrete to chilled brood only. A very positive sign.
- In the later stages a black scale is formed in the cell which is easily removed by the bees.

5.2 The signs of the bacterial diseases American Foul Brood (A.F.B.) and European Foul Brood (E.F.B.) and describe their effect upon the colony.

5.2.1 Field diagnosis of AFB and EFB and the signs of the two diseases.

ADAS leaflet # P306 'Foul brood of bees: recognition and control' should be obtained (free of charge) from your Regional Bee Inspector or through your Branch Secretary.

- Both diseases are diseases of the brood and there are no signs associated with the adult bees in an infected colony. In order to diagnose either in the field, it is necessary to open up the colony and examine the combs containing brood. To do this properly it is necessary to shake the bees off the comb before examining it, leaving no more than a few bees on the comb. The reason for this is that in the early stages only an odd cell or two will be exhibiting the tell-tale signs. This important aspect of searching for the diseases is frequently overlooked and inadequately expressed in much of the literature. There is a right and a wrong way of shaking bees off combs, the objective is to rid the comb of bees and keep them in the hive (not flying around the apiary); therefore raise the comb slightly and shake it sharply in the brood chamber without jarring the rest of the colony.

5.2.2 In order to diagnose the diseases in the field it is easier to remember the signs if one has an understanding of the progress of the diseases:

• AFB (American Foul Brood): Causative agent - *'Paenibacillus larvae larvae (P.l.larvae)'*. The larva is fed the AFB spores with the larval food. The spores germinate in the ventriculus and the larva dies after the cell is sealed. The germinating spores break through the wall of the ventriculus into the haemolymph and the larva dies of septicaemia * ; then the whole larval form disintegrates, melts down, becomes thick and sticky and finally dries to a hard scale on the lower angle of the cell. During this deathly saga the colour changes from white to black. It is most important to note that prior to the sealing of the cell, the larvae appear to be perfectly healthy.

* Septicaemia - is the circulation and multiplication of micro-organisms in the blood.

• EFB (European Foul Brood): Causative agent - *Melissococcus pluton*. The larva is again fed the pathogen, this time a bacteria which is not spore forming as was AFB, which multiplies in the ventriculus by using the larval food and the larva dies before the cell is sealed due to starvation. It dies at about day 3 or 4, so it is quite large when it dies. A dead larva is not sealed by the bees and is removed. During the starvation period the larva contorts into unnatural shapes in its cell and changes colour from a pearly white to cream to yellow to light browny green (colours are difficult to describe in words; any deviation from the pearly shiny white must be regarded with suspicion). When the bees remove the dead larvae, they are removed in one piece and they are either there to see or else the signs have been removed by the bees.

5.2.3 Signs of AFB caused by the spore forming bacteria *'Paenibacillus larvae larvae'*:

a) open brood - no signs,
b) sealed brood, many signs as follows:

- After the larva dies, the domed cells become moist and darken in colour.
- Cappings then sink and become concave (still moist and discoloured).
- Holes appear in the cappings (ie. perforated).
- Matchstick pushed through sunken capping to test for roping of the contents. Length of 'rope' between 1 and 2 cms. This roping is considered to be a positive identification of the disease. Colour of cell deposit is from light brown to nearly black. The roping test can only be done between the time the larva has 'melted' and the melt thickened slightly and before the remains of the larva has dried.
- The remains dry out on the lower angle of the cell and form a hard black scale. By the time the scale is formed, the bees have uncapped the cell completely and tried to remove the scale. In order to see the scale the comb must be held at an angle with the top bar closest to you and the bottom of the frame away from the body (angle about 45° to the vertical). Good light is essential, some books say from the back while others say from the front; we think either is acceptable depending on whether you are in or outdoors. In the early stages of the infection, possibly only one or two cells may have scale and this is why it is so important to clear the frames of all the bees when doing an inspection for foul brood.
- Brood combs which have a 'pepperpot' appearance (ie. empty cells among sealed brood)

108

should be treated with suspicion and examined closely for any sign of scale.

c) AFB infections have no smell; many books indicate a foul odour. Bacillus larvae when sporulating releases an antibiotic preventing any secondary infections. If an offensive smell is present it will be due to secondary infections of some other cause or the confusion may arise because when the bacteria are in the rod form all the cells are sealed and no odour can be released. Rely on visual signs, not odour, for AFB.

d) AFB is easily identified visually in the field. However, it can be confirmed if necessary by laboratory tests usually on a piece of scale from the comb.

5.2.4 Signs of EFB caused by the non-spore forming bacteria '*Melissococcus pluton*':

a) Sealed brood - no signs (the larva dies before sealing).

b) Open brood - the signs are as follows:
- Larvae are usually in unnatural contorted positions in the cells; twisted spirally or flattened out lengthwise (nb. stomach ache is a good analogy).
- The colour changes from a pearly white of a healthy larva to dull cream, to light brown and eventually a greeny hue. The colour change should be associated with the unnatural positions.
- The dead larvae have a melted down appearance but still have a larvae-like shape.
- Again EFB does not itself smell. However very often an offensive smell is present on combs with EFB infected larvae; these are secondary infections often associated with EFB and are another indication that the disease may be present (2 common secondary infections are *Bacillus alvei* and *Bacterium eurydice*).

c) EFB is very difficult to diagnose positively in the field for the following reasons:
- The larvae are removed quickly from the hive once they are dead so the evidence is often removed and not there for the beekeeper to see.
- Any diseased larvae can be confused with other brood diseases, such as Sac brood or Neglected drone brood, unless the beekeeper is very experienced. Most Bee Inspectors will remove a frame with dead and dying larvae for laboratory analysis.
- The best time to look for EFB is when the brood outnumbers the adult bees in the spring about mid April to early May. At this time the chances of spotting the diseased larvae are greater because the house bees are fully stretched under these conditions.

5.2.5 The effects of AFB and EFB on the colony.

• AFB. Unlike EFB there is no apparent seasonal outbreak and it can occur at any time. If the number of infected larvae are less than 100 the colony may recover but if the number of cells is greater than 100 then the colony is likely to succumb. The colony that becomes badly infected gradually becomes weaker until its demise; the signs are increased pepperpot appearance of the combs and fewer and fewer worker bees. The danger now exists of the weak colony being a prime target for robbers. Finally the colony either dies out naturally or is robbed out thereby spreading the disease to another colony.

• EFB. Little is known about why EFB suddenly appears in a colony and then disappears just as quickly; a characteristic of strong colonies. Weak colonies with EFB fail to recover and will eventually die out. It is believed that EFB is now endemic in the United Kingdom and the beekeeper seldom sees the signs of the disease except when there is a sudden start to a honey flow and many young bees are recruited to foraging duties or during the spring build up when the amount of brood is greater than the adult bee population. EFB is now regarded as one of the bee diseases that are occasioned by stress.

Both AFB and EFB are notifiable diseases. As soon as possible the beekeeper should get in touch with the local Bee Inspector to ask for confirmation of the disease. It is possible to have both AFB and EFB present in the same colony.

5.3 Methods for detecting and monitoring the presence *of Varroa Destructor* (a mite) and describe its effect on the colony.

5.3.1 Methods for detecting the presence of *Varroa Destructor.*

Varroa Destructor is a parasitic mite which lives on the exoskeleton of the honeybee and breeds in the sealed cells of worker and drone brood. All methods of detecting the mite are visual. There are now a wide number of methods for detecting *Varroa Destructor* infested colonies but for the purpose of this section only three simple methods will be described which are seasonal in use.

a) Examination of hive debris on the floorboard in early spring.
This method has been advocated by the NBU for some years and at the time of writing they will still undertake the examination free of charge. The method has its limitations and is not wholly reliable particularly when the infestation in the colony is light. The debris from all the floorboards in an apiary is collected and bundled together and sent to the NBU with the name and address of the beekeeper and his apiary. After examination, the NBU send back a report stating the results. If the sample comes back with a negative result and a light infestation has been missed it will be of little consequence on colony performance during the coming season. The method must be quite attractive to many beekeepers until such times as it is discontinued or until the apiary concerned has been diagnosed positive.

b) Uncapping brood.
This is used particularly on drone brood at the pink eye stage of development, using an uncapping fork to expose the drone pupae during regular colony inspections. Care must be taken when examining the larvae for reddish coloured mites; in the early stages of development the nymphs are virtually colourless and translucent. Special frames of drone foundation can be inserted in the brood chamber for this purpose. Note that this is also a manipulative method for the control of Varroosis without chemicals.

c) Bayvarol/Apistan test using a varroa screen.
The varroa screen must be used in conjunction with a paper insert below together with one strip of Bayvarol manufactured by Bayer or Apistan made by Sandoz. It is prudent to smear a thin layer of grease around the edges of the paper insert to prevent any mites that are knocked down and not killed from walking back into the hive again. Bayvarol and Apistan are the only medicaments approved for use in the UK at the time of writing. Bayvarol strips contains 3.5mg of flumethrin and Apistan strips contain 8g of fluvalinate, the test strip should be left in the colony for 24 to 48 hours. The strip and insert are then removed and the insert examined with a magnifying glass for dead mites. The strip should be inserted between the centre frames of the

cluster and can be introduced without moving the frames and disturbing the colony. It is the most effective detection method available at the present time. The strip can be re-used to test other colonies in the same apiary, noting that other diseases can be transferred from one colony to another on the same strip. When Bayvarol is used for treatment the strips are left in the colony for 5 to 6 weeks, ie. the strips have a working life, according to Bayer, of 6 weeks using 4 strips in an average size colony. For testing purposes it would be prudent to use the single testing strip in the colonies for say no longer than 21 to 30 days of actual use before it is scrapped thus ensuring a reasonable surface density of flumethrin to knock down any mites that may be present. The test can be used at any time but is usually applied in the spring and autumn. Only 2 strips of Apistan are required to treat an average size colony.

5.3.2 Methods for monitoring the mite level in a colony throughout the year.

Once the mite has been detected in the colonies of an apiary, any monitoring process must involve counting dead mites that have died a natural death or counting the mites killed by knock down tests at regular intervals. If mites have been detected in one colony in an apiary, then it is safe to assume that all colonies in that apiary are infested or will be very shortly even if tests on these other colonies prove negative in the initial stages.

To monitor dead mites it is necessary to have a special floorboard with an integral Varroa screen (which are unpopular because of their high cost) or resort to a DIY job with the existing floorboard turned through 180° with a screen above.

The frequency of each monitoring session must be decided and the more often the better for the best results. Whether this is once a week or once a month must rest with the beekeeper and the amount of time that he can devote to the monitoring as part of his management system. Once the number of colonies rise say above five the work of monitoring will quickly become very time consuming and tedious.

It is not much use counting mites as a continual monitoring process unless action is taken if the levels rise above a predetermined threshold. For management purposes it is not clear what this level should be. It should also be noted that if the colony has had supers added then it would be unwise to treat until they are removed which is likely to be at the end of the season. The question must therefore be asked whether there is any purpose in monitoring on a continuous basis throughout the season?

We are of the opinion that any monitoring should be regarded, at present, as experimental in order to determine possible management methods for the future. The equipment required to provide continuous monitoring demands screens that do not corrode and are made of a material that cannot be damaged by the bees themselves.

It is highly unlikely that bee farmers with commercial enterprises will have sufficient time or effort available to undertake monitoring programmes; the economics of running such businesses would preclude it. Most of the hobbyist beekeepers are unlikely to be interested also. Thus any monitoring is likely to be left to the informed hobbyist beekeeper who has an interest in this new disease. Thus the majority of beekeepers in the UK will require control methods to be applied annually or bi-annually that are simple to use and relatively cheap.

During 1998 the CSL produced a calculator (based on work undertaken by Dr.Martin) which takes into account the natural mite drop, the month of the year the mite drop is measured and then predicts whether to treat or how many days grace the beekeeper has before treatment is necessary. It is not clear whether it has found much favour among beekeepers and how effective it is as a management tool. Time will tell.

5.3.3 The effects of the mite infestation on the colony.

It is instructive to examine the effects of the parasite on the individual worker bee in the colony:

The number of mites per cell (or per bee) has a marked effect on the haemolymph of the bee, the protein content being reduced by 15 to 50%. Brenda Ball quotes 1 to 3 mites cause a 27% reduction and 4 to 6 mites a 50% reduction.

The protein reduction in turn results in a marked reduction in the final weight of the bee (6 to 25 % weight loss) and a reduction in the longevity of about 34% to 68% of the adult life span.

If 5 or more mites are present then there is a high probability that the bee will be killed. If not there will be marked damage to the wings, legs and abdomen. Infestation must be at an advanced stage and the visible signs such as crawling and deformed bees at the entrance will only represent a very small fraction of the colony damage.

The morphological changes are small when bees are infested with 1 or 2 mites (eg. 1 to 3% reduction in wing length). This demonstrates the importance of early detection in the management of colonies because it is so difficult to detect any damage.

It takes approximately 3 to 5 years before the colony is weakened, that is, when the mite population = 30 to 40% of the adult bee population (eg. 40,000 bees and 12,000 mites). At this stage there will be a rapid decline in the adult bee population with severe brood damage, reminiscent of the signs of EFB. The death of the colony will quickly follow.

If the infestation reaches a level of 1 mite / cell and if treatment is not undertaken the colony will die out in 2 to 4 years. At the terminal stage the collapse is very rapid.

It will be clear that due to the general weakening of the bees and the colony coupled with the reduction in longevity, the normal house bee and nurse bee duties will be seriously disrupted leading to poor colony hygiene and a deterioration in the hive environment.

It is clear from the above that the signs of Varroosis are virtually impossible to detect visually until it is too late. Early diagnosis is essential for good bee husbandry.

Varroosis is also a notifiable disease. Most of England is now recognised as an infested area. Southern Scotland is now infested despite restrictions on moving bees. These restrictions are still in force in Scotland at the present time.

112

5.4 The effect of *Acarapis* (a mite) and *Nosema* (a protozoa) upon the colony.

These pathogens are the cause of two adult bee diseases namely Acarine and Nosema; both are endemic in the United Kingdom just as Varroosis is now endemic. For completeness the other adult diseases are:

Amoeba,
Dysentery (not strictly a disease but a condition),
Varroosis (used to be called Varroasis and incorrectly called Varroa),
Viral diseases (eg. Paralysis and others).

5.4.1 Acarine- Causative agent - *Acarapis woodi* (Rennie) - a mite in the class Arachnida.

• These mites (c. 150μm x 65μm requiring a microscope to see them) were discovered by Dr. Rennie at Aberdeen University as a research project funded by the philanthropist Mr. Wood. It is usual for a new biological discovery to be partially named after the scientist who did the research work. In this case both the scientist and the philanthropist are named. This work on Acarine (called at that time 'the Isle of Wight Disease') was commissioned in 1921 after many colonies had been wiped out in UK.
• The EEC terminology for Acarine is acariosis and there were proposals, in 1990, that it should become a notifiable disease. At the date of writing nothing more seems to have been heard about this absurd proposal.
• The female mite enters the first thoracic spiracle and then into the associated trachea where it lays its eggs and feeds on the haemolymph of the bee by piercing the trachea walls. The trachea can become completely blocked by the mites and their offspring which can only be seen by dissection and examination under a microscope.
• Signs of the disease: Despite the large number of references to the signs of Acarine in beekeeping literature, Dr. Bailey's work has shown that there are no visible external signs of this disease. The disease has no effect on the flying ability of the bee but it does shorten its life slightly (time not quantified).
• The following signs are those of Chronic Bee Paralysis Virus (CBPV) often confused with Acarine:

1. Bees crawling with fluttering wings.
2. Bees clinging to plants or blades of grass near the hive.
3. Bees with a shiny exoskeleton and bloated abdomens.
4. Crawling bees outside the entrance to the hive, may be in large numbers.
5. Bees with dislocated or partially spread wings (K-wings).
6. Bees huddled together on the top bars or on top of the cluster in the hive (these bees do not move away from smoke).

• Diagnosis of Acarine can only be confirmed by dissection and microscopic examination of the first thoracic trachea. When the disease is present the trachea will be discoloured and not the normal creamy colour of healthy adult bees. The trachea can be infested either on one or both sides.

• It should be noted that there is no correlation between the *Acarapis woodi* and CBPV and *Acarapis woodi* has not been proved as a vector for the spread of CBPV. This is very curious because when crawling bees, etc. are found in a colony and the bees are examined, in a large percentage of the cases Acarine will be present.

• Treatment: All the old treatments for Acarine such as Frow Mixture, Folbex (Chlorobenzilate) or Folbex VA (Bromopropylate) have been banned by the EEC on the grounds that they are all carcinogenic thus leaving the United Kingdom without an approved medicament. However, our experiments in Devon have shown that the approved treatment for Varroosis, namely Bayvarol or Apistan, is effective.

• The effect of Acarine on the colony during the active season appears to be minimal and in a good season with good honey flows can disappear completely. The disease takes its toll if the colony goes into winter and the disease is not treated; something which is unlikely to happen these days when every beekeeper is treating his colonies with Bayvarol/Apistan during August each year.

5.4.2 Nosema - Causative agent - *Nosema apis* (Zander) - a spore-forming protozoa.

• The Nosema spore (6 to 8μm) can only be observed using a compound microscope. It was discovered by Prof. Enoch Zander at Erlangen in the early part of the century. The protozoa multiply in the ventriculus (gut) of the honeybee (30 to 50 million spores when infection fully developed) and impair the digestion of pollen thereby shortening the life of the bee. It does not affect the honeybee larvae.

• Signs and effect:
    1. Infected bees themselves show no outward signs of the disease.
    2. Colonies fail to build up normally in the spring (see Appendix 3).
    3. Badly infected colonies in the early part of the year:
        - exhibit signs of dysentery (soiled combs and soiled entrance),
        - dead bees outside hive entrance (after cleansing flights).

• Diagnosis of nosema can only be confirmed by microscopic examination.

• Treatment:
    a) Fumidil 'B' inhibits the spores reproducing in the ventriculus. It does not kill the spores.
    b) Autumn treatment: Fumidil 'B', followed by spring treatment the following year.
    c) Spring treatment: Bailey frame change plus Fumidil 'B' administered in syrup. Dosage 166mg of Fumidil 'B' to one gallon per colony.
    d) Good beekeeping practices prevents spread of infection in both the hive and the apiary eg. no squashing of bees during manipulations and prevention of robbing, drifting, etc.
    e) Disinfection of infected comb and hive parts with 80% acetic acid (100ml/brood box for one week). Note that acetic acid should be placed on top of the frames because the fumes are heavier than air and sink to the bottom between the frames. Acetic acid is very corrosive and metal ends should be removed from frames and any remaining metal work should be greased before treatment.

• Other points:
    a) Due to the high incidence of nosema in UK it is virtually essential to monitor twice a year by taking samples in spring and autumn and treating as required.
    b) It is understood that using Fumidil 'B' regularly as a prophylactic for nosema is unlikely, but not

impossible, to produce forms of *nosema apis* resistant to the antibiotic Fumidil 'B' (Prof. L.Heath).
c) It is important to ensure that any Fumidil 'B' used for this treatment is not time expired and has been stored in a cool dark place.

5.4.3 Amoeba (*Malpighamoeba mellificae*) a protozoan amoeba-like parasite which ultimately encysts in the malpighian tubules.

• The cyst which is found in the malpighian tubules is c. 10 to 12μm and can only be detected using a compound microscope. The cysts germinate, develop and multiply in the ventriculus (gut). The amoeba then make their way into the tubules and eventually form cysts which pass into the small intestine and rectum and are voided in the faeces. The infection seems to have no effect on the colony.
• Signs: There are no external signs.
• Treatment: No medicaments are available for treatment and putting the colony on to clean comb, as for nosema, and disinfecting the comb and hive parts is the only treatment available.

5.5 Ways of controlling Varroosis using one registered product and one recognised biotechnical method.

5.5.1 Control using a registered product.

Bayvarol and Apistan are the only treatments licensed for use in the UK at the time of writing. All medicines are normally licensed 'P' category (ie. for sale through a pharmacy). The 'P' category is broken down further into 'POM' (ie. only available on prescription from a veterinarian, doctor or dentist) and 'PML' (available from pharmacy and merchants list) which can be sold through licensed agricultural merchants (as in the case of Bayvarol and Apistan). If it was licensed 'GSL' (general sales list) it could be sold anywhere. It is interesting to compare the licensing of Fumidil 'B' which can go directly into the food chain and is in category 'GSL'. Fumidil 'B' should not be used when the honey supers are on the stocks.

The active medicament of Bayvarol is flumethrin ($C_{28}H_{22}CL_2FNO_3$), a synthetic pyrethroid, contained in a polythene strip. The manufacturing process involves coating very tiny beads of polythene with flumethrin and then moulding them into a strip. No plasticizers are used in the moulding process consequently there is no migration of the active material from the inside to the outer surface as the surface flumethrin is used up by contact with the bees. The initial surface density of flumethrin strip is 500μg per gram and each strip contains 3.5mg of flumethrin when new.

The concentration of flumethrin must be sufficient to kill the mite and low enough not to kill the bees. At low levels of concentration with a low mite kill there are a large number of mites which can breed a resistance to the medicament. This demonstrates the necessity of not re-using strips having a low concentration for longer periods.

Flumethrin is a nerve poison. It is believed to enter the mite by absorption through the soft pads at the ends of its legs. The toxin acts on the pre-synaptic sodium and potassium channels in the nerve, increasing the number of action potential impulses in the insect nerve leading to a lack of co-ordination, loss of normal bodily functions and death presumably by starvation.

The treatment involves a dosage of 4 strips per colony (total 14mg) for a period of 6 weeks. We consider that the best time to treat is in the late summer / autumn when all the supers have been removed from the hives to minimise residues in honey and wax in the supers. However, if signs of Varroosis are evident in a colony, such as deformed bees, the time to treat is immediately.

We have discussed the use of having to wear gloves when using this product as the Bayer film about Varroa shows the operator wearing industrial heavy duty gloves. It has been considered by the registration authorities and it is not necessary to wear them due to the low levels of flumethrin on the strips, the low dermal absorption and its overall low toxicity. However, hands should be washed after handling the strips before eating. Additionally, there are people who have pyrethroid allergies and they will of course have to handle the strips with great care.

5.5.2 Treatment by a recognised biotechnical method.

Removal of sealed drone comb. This must be the most popular method and used by most beekeepers as it is simple, quick and fits in well with the routine colony inspections. A frame of drone comb is put into the brood nest usually around April time, when the queen is laying well, and removed when the brood is sealed. A further comb is put in and the operation repeated. It relies on the preference of the mite to breed in drone brood.

There are two ways of dealing with the drone comb complete with drone larvae or pupae plus mites. One way is to decap and hang it up for the blue tits to clean out. We do not regard this as a good idea as other bees can be attracted to it and re-infestation is possible unless the comb has been in the freezer to kill both larvae and mites. The other way is to uncap and hose the larvae and mites from the cells with a jet of water.

Instead of a special brood frame containing drone comb a super frame can be put into the middle of the brood nest and the bees will build drone comb on the bottom of the frame. When sealed with drone brood it can be sliced off and replaced to start the process all over again.

The beekeeper may wish to consider whether this method which wastes a lot of the colony's effort is worth employing; we, ourselves, do not use it under normal circumstances.

5.6 Awareness that *Braula coeca* is neither a mite nor a parasite, but is an insect that steals food.

Braula coeca (Nitzch) in the family *Braulidae* is called a bee louse but in effect it is a wingless fly and not a true louse. The family is contained in the order *Diptera* (true flies). The two true lice orders are *Anoplura* (sucking lice) and *Mallophaga* (biting lice). The relationship between *Braulidae* and other *Diptera* is still enigmatic and discussion continues about the placement of *Braulidae* in the animal kingdom. All adult *Braulidae* lack halteres and wings.

All *Braula coeca* are reported to be inquiline (an animal living in the home of another animal) and none are reported to be harmful to the colony. The life cycle is as follows:

- Eggs are laid on the inner side of honey cappings and sometimes on the wall of cells filled with honey.
- Eggs hatch to larvae which feed on wax and pollen (found in their intestines) forming ever-lengthening tunnels in the wax cappings which widen as the larvae grow.
- The larvae pupate at the wide end of the tunnels and finally emerge as adults.
- At emergence the adult *B.coeca* is white and changes to its characteristic reddish / brown colour in 12 hours as its exoskeleton hardens.
- Development is entirely under the cappings of honey cells and it is not associated with brood cells in any way.
- Adults mainly inhabit the petiole of worker bees, queens and occasionally drones. They move to the mouthparts of the bee when it starts to feed.
- When the bee is feeding the *B.coeca* resides on the open mandibles and labium reaching into the cavity at the base of the extended glossa near the opening of the duct of the salivary glands. It is not clear whether it feeds on the salivary gland secretions but it is thought to be highly likely.
- The queen appears to be more attractive than the other castes and the maximum number on a queen has been reported as 30 at one time. The maximum daily collection from a queen is reported as 371! Some authorities consider that infestation at these levels would have some effect on reducing the laying ability of a queen.
- Breeding takes place between May and September and the louse has the ability to overwinter with the bee.

No ill effects on the colony have ever been reported but damage to cappings of honey sections and cut comb honey from the larval tunnels occurs. It is always advisable to put prepared comb for sale into the freezer (-15⁰ C) for a couple of days to kill any eggs and larva. It is seldom necessary to specifically treat a colony for Braula infestation. It is likely that all chemical treatments for Varroosis will be lethal to the *Braula coeca* and it could possibly become an 'endangered' species.

5.7 Ability to distinguish between *Varroa Destructor* and *Braula coeca.*

All detection for Varroosis is dependent on being able to recognise a *Varroa Destructor* mite and not to confuse it with a *Braula coeca*. The physical differences are as follows:

Braula coeca: ellipse shaped c.1 - 2mm with 6 legs, coloured reddish brown. It is a wingless fly. Initially it is white and takes about 12 hours from hatching to develop its colour. The head and posterior end of its abdomen are on the ends of the major axis of the ellipse, the legs are on the sides associated with the ends of the minor axis of the ellipse looking down on the dorsal side. Easily seen by eye riding on worker bees and very often the queen is infested. Causes no harm to queen or bees but the larvae spoil honey comb with fine tunnels in the cappings.

Varroa Destructor: also ellipse shaped c. 1.1 - 1.7mm. with 8 legs, coloured reddish brown the same as the Braula coeca. The legs are on the ventral side and cannot be seen when it is viewed looking down on the dorsal side. It travels 'blunt end first', its legs being on the sides associated with the ends of the major axis. It is an arachnida and is in the spider class in the animal kingdom not the insect class. These mites are difficult to detect as they feed on the haemolymph by piercing the membrane between the abdominal segments on the adult bee and breed in the capped brood cells.

5.8 The current legislation regarding notifiable diseases of honeybees.

There are 3 major pieces of legislation as follows:

> The Bees Act 1980.
> The Bee Diseases Control Order 1982, S.I.107 (AFB, EFB, Varroasis).
> The Importation of Bees Order 1980.

AFB, EFB, Varroosis: the major points in the legislation are as follows:

a) Notification of disease: the beekeeper who suspects disease shall notify with all speed and shall not move bees, hive, etc. until the authorised person, the Regional Bee Inspector (RBI), has examined them.

b) Precautions against spread of infection:
- Authorised person may take samples of combs (for AFB, EFB) and combs plus bees and debris (for Varroosis).
- Authorised person, if he suspects disease, shall serve a notice prohibiting removal except by licence ie. a stand-still order on apiary/s.
- Authorised person may mark any hive or appliance for identification purposes.
- If samples are positive, the authorised person shall serve a notice.

• AFB:

a) If the authorised person and the beekeeper agree the disease is present, beekeeper signifies his agreement by signing the notice.

b) Notice requires destruction, details of which are specified.

c) Destruction to be supervised.

d) Notice in force until date subsequently notified to the beekeeper by RBI.

• EFB:

a) Same as AFB except colony may be treated in specified manner in lieu of destruction but again under supervision.

b) The authorised person may serve notice on any other beekeeper whose bees may be likely to come into contact with diseased bees.

• Varroosis:

a) Same as EFB but also require isolation period as stated in the notice.

b) Declare an area as infected by being published publicly (in the local press?).

c) Beekeepers are required to:
- Notify if disease is suspected.
- Must not interfere with identity marks on equipment.
- Provide facilities and information to authorised person.
- Must not treat bees with drugs that disguise presence of disease.

It is advisable to obtain a copy of the legislation which is obtainable from HMSO. Note that 'shall' in legal terms is mandatory, whereas 'may' is not mandatory.

The Importation of Bees Order 1980 prohibits the importation of bees but makes provision for the importation of queens and attendants from Varroa free countries under licence. In practice this is now New Zealand only; and some authorities argue that because of the lack of knowledge about Kashmir Bee Virus, this country should also be banned. At time of writing the importation of package bees is under consideration by DEFRA. Note that the order does not prohibit the movement of genetic material such as eggs or sperm.

The statutory requirements in respect of Varroosis have not yet been repealed and it unlikely that prosecution would follow if an infestation of varroa mites was not reported. The whole of England and Wales is regarded as an infested area though there may still be some isolated pockets of disease-free bees.

5.9 The national and local facilities which exist to verify disease and advise on treatment.

5.9.1 The expert services available to the beekeeper at national level.

At national level the following organisations can be contacted in the event of information being required on diseases:

•National Beekeeping Adviser, Central Science Laboratory, National Bee Unit (NBU), Room 10GA05, Sand Hutton, Nr. York YO41 1LZ, telephone 01904 462510, fax 01904 462240, E-mail: m.bew@csl.gov.uk - for analysis of samples for all diseases and poisoning incidents. It should be noted that charges are levied for most of their services and these charges should be ascertained before entering into a contract with MAFF for any services required.

•International Bee Research Association (IBRA), 16 North Road, Cardiff; CF1 3DY, telephone 01222 372409, fax 01222 665522, E-mail: ibra@cardiff.ac.uk - at the time of writing an extensive library and many publications for sale are available.

•British Beekeepers' Association (BBKA), BBKA Headquarters, NAC, Stoneleigh Park, Warwickshire CV8 2LZ, telephone 01203 696679, fax 01203 690682, E-mail: s.edwards@bbka.demon.co.uk - for general guidance. BBKA can provide many useful publications and also provides initial advice on legal matters.

5.9.2 The expert services available to the beekeeper at local level.

At local level, advice and assistance is available from various organisations depending on which county you reside in:

a) County Beekeeping Association (contact the Secretary).
b) District or Branch Association (usually the first person to contact would be your own Secretary) assuming that you are a paid up member.
c) Your local DEFRA offices (to obtain assistance for suspected foul brood infection). It

should be noted that most counties only maintain this assistance on a part time basis from April to October. There is no charge levied for suspected foul brood inspections by these officers.

d) Regional Bee Inspectors who are available throughout the year.

e) Agricultural Colleges can often give assistance. Most of the CBIs were at one time based at such a college.

f) County Bee Instructors (CBI) or Lecturers (CBL) will give advice if they are available. Most, if not all, of the posts are now abandoned and, if they do exist, are part time only.

g) The secretaries of most district and branch associations are in a position to provide addresses and telephone numbers as they receive (or should receive) all the up to date information.

h) Many local beekeeping branches provide their own diagnostic facilities for adult bee diseases and advice on treatment, generally at no cost to their members.

5.10 Where to obtain assistance if any poisoning by toxic chemicals is suspected.

As legal action is sometimes involved when poisoning of honeybees by toxic chemicals occurs, we consider it better to provide a fuller account of the subject and identify what action to take in such cases and the practical measures possible when prior notification is received.

• The main problems are caused by agricultural spraying of pesticides (a generic name for insecticides, fungicides, herbicides, etc.) for a variety of reasons to combat damage to the crop and hence procure a greater yield for the grower.

• Some growers and farmers undertake the spraying of their own crops, others retain the services of professional spraying organisations. Spraying is a skilled job. Troubles associated with bee fatalities only occur when inexperienced and untrained staff are left unsupervised or the operators take short cuts and/or do not follow the makers' instructions. There is now legislation requiring the owner of the crop to provide the beekeeper with a minimum of 48 hours notice if he has colonies nearby and which are likely to be affected (Control of Pesticides Regulations 1986).

5.10.1 Bees and brood can be killed by toxic chemicals in three ways:

- By direct contact (through the integument).
- By eating (into the alimentary tract).
- By breathing (fumigation into the trachea via spiracles).

5.10.2 Contact with the poison can occur in three ways:

- By direct contact on flowers that the bees are working, which has accidentally been sprayed with the treated crop (eg. weeds in the hedgerows). Note that the treated crop may not necessarily be in bloom.
- By being caught in the spray on the crop the bees are working.
- By flying over a crop which is being sprayed.

120

5.10.3 Spraying can be done in three ways

 - By fixed wing aircraft (which is the worst - minimum control over the spray).
 - By helicopter (more controlled, note down draught).
 - By tractor (which is the least damaging - ie. to working bees).

Fruit growers spray the most and cause the least damage to bees. The worst crops for spray damage to bees are *Cruciferae* (eg. rape) and field beans. Note that field beans are often sprayed while they are in the early stages of flowering to combat aphid infestation; honeybees are not working the crop at this stage but it is likely to kill off all the bumblebees and therefore ruin a possible honey crop for the honeybees. The authors have found some farmers unaware of the damage they have done in this respect (bumble bees pierce the base of the bean flower with their strong mandibles allowing the honeybee easy access to the available nectar).

5.10.4 The time that spraying is actually carried out is very important; this is related to the times that honeybees are expected to be flying:

 - Before 8 am and after 8 pm are the best times.
 - During the day is the worst time irrespective of the weather conditions.

5.10.5 Diagnosis of spray poisoning:

 - Can be easily confused with CBPV and starvation. Piles of dead bees outside the hive or shivering, staggering and crawling bees also outside the hive.
 - Only laboratory tests and analysis provide a satisfactory answer. Note that a large sample is required because many tests often have to be performed on a large number of small samples to trace a particular pesticide (there are many hundreds of different types). This is the reason for obtaining as much detail as possible about a spray incident, in order to make the identification of the poison as easy as possible.
 - A few or many bees die suddenly depending on the poison and how much has been taken into the colony or how many foragers have been affected.
 - The number of foragers at the entrance is less than normal.
 - Poisoned bees from inside the hive are ejected.
 - Colony tends to become bad tempered.
 - The crawling bees tend to have curled up abdomens.
 - Returning foragers spin around on the ground until they die.
 - Generally there are many dead or dying bees in front of a colony that has suffered poisoning.
 - Dead bees usually have their proboscis extended.
 - Honey is not usually affected. Poisoned bees are not admitted into the hive and therefore not unloaded by the house bees. Thus, there is no food transfer to other workers providing an automatic protective mechanism.

5.10.6 Action to be taken when spray poisoning is suspected:
a) Comply with the agreed procedure of your local spray warning scheme.
b) Record as much detail as possible about the incident because if litigation is involved it will be

some considerable time in the future.

c) Photographs of the colonies and the sprayed crop are often overlooked and are extremely useful at a later date. Place a newspaper with the date at the side of the photograph shot.

d) A large sample is required for reasons outlined above; BBKA Advisory Pamphlet # 27 advises 3 samples each of c. 300 bees to be sent to the NBU at Sand Hutton with the Spray Incident Report which should include the following details:

- Time and date discovered.
- The number of hives affected plus observations on each.
- Estimate of dead bees from each hive(c. 5,000 bees weigh 1 lb or 454g).
- Condition of bees and colour of the pollen sample from dead bees.
- Behaviour of colonies (eg. temper, bees being ejected, etc.).
- Sketch map of area and OS grid references showing apiary and crop (don't forget to mark North).
- Weather conditions (wind speed and direction, temperature, rain/fine/sunny/etc.).
- Discuss with crop owner and seek confirmation of spraying and the spray used. Visit site with owner, if possible, and determine crop acreage and weeds treated.
- Determine method and time of application together with the flowering state of the crop (nb. photograph of crop).
-Names, addresses and telephone numbers of all concerned.

e) It is important to advise your Branch Sec. and/or your Spray Liaison Scheme representative in order that they may alert other beekeepers in the same area.

f) Don't forget to label the sample of poisoned bees in your freezer in case any of the samples are mislaid.

5.10.7 First aid treatment for stocks which have lost their foragers through poisoning is to give a gallon of syrup (50:50) as soon as practicable

5.10.8 Practical measures to be taken when prior notice of spraying is received:

• It is now mandatory that 48 hours notice must be given to the beekeeper of any spraying operation.

• If it is likely that the colonies will suffer then:

- The colonies should be moved to a safe place if possible; this means to a site at least 3 miles away to ensure that there is no possibility of them returning to the crop.
- If the colonies cannot be moved, it will be necessary to confine the bees for 24 hours maximum (denying them access to the crop); this is the maximum time that bees can safely be confined providing precautions are taken.
- The colonies must be prevented from over-heating; therefore additional comb should be given together with an empty eke on top with a large sponge soaked in water suspended in it.
- The colony should be closed up at night.
- Colonies should then be kept in the dark (tenting with black polythene or covered in straw if

available).

- Colonies can be closed up with crushed ice from a refrigerator; it is claimed that the bees will not pass the entrance even when there are holes in the ice blockade.

There is a fair amount of work preparing colonies that are to be closed up for 24 hours and moving them would probably amount to the same effort and be safer, particularly if the spray used was lethal for more than 24 hours. In the authors' opinion it is safer to move colonies rather than close them, certainly if the colonies are strong and if the weather is warm.

- Note the BBKA spray liaison scheme which is understood to be working well in some counties but is lacking in others.
- The most important factor is for the beekeeper to develop a good working liaison with the local farmers and their spray contractors. In the authors' experience, the farmers and growers are extremely co-operative and welcome any liaison but the beekeeper has to make the running. If he does, it usually gets to the stage where the farmer or the spray contractor telephones the beekeeper to discuss a spraying operation before it starts; a good reason for having your telephone number displayed on the apiary hives.

5.10.9 Where to obtain assistance in suspected cases of spray poisoning.

These incidents are all handled by the NBU in Sand Hutton in the initial stages; it would be wise to telephone them first to check that the sample should be sent direct to them. They in turn alert a local MAFF official who will visit the site and prepare an incidence report. They will be the body who will prosecute the sprayer if they consider he has been remiss.

If you are a member of your county spray liaison scheme then you will have procedures which are discrete to your county scheme.

5.11 How comb can be stored to prevent wax moth damage.

5.11.1 Wax moth damage to stored comb.

5.11.1.1 Greater wax moth (Galleria mellonella).

The adults have a wing span of 1 to $1\frac{1}{2}$ inches (25 to 38mm) and enter the hives at night to lay eggs. The eggs hatch to larvae and when fully grown they are about $\frac{7}{8}$ inch (22mm) long and quite distinctive with a dark head. The larvae pupate, usually in a boat-shaped groove chewed into the woodwork of a frame, the chrysalis eventually hatching to the adult form. The damage is caused during the development of the larval form. The larva has the ability to digest wax but it also needs protein which is obtained from pollen and larval debris of the honeybee.

The life cycle is approximately: egg - 7 days, larva - 15 days, pupa - 28/32 days. The times are very variable and depend on temperature (egg to adult on average is 50 to 54 days).

In warm weather there is the possibility of all the comb in a full brood chamber being turned to dust in about 14 days and much of the woodwork damaged if it is off the hive with no bees. A strong colony will not tolerate the moth and keeps itself in a healthy state and no damage is caused. Weak colonies can be damaged with the bees in occupation.

These moths are generally only troublesome in UK when the comb is not in use, they can however be very real pests in tropical climates.

It is possible to introduce bacteriological control by impregnating the stored comb with spores of *Bacillus thuringiensis* which kills the wax moth larvae. This form of control has been used in USA but is not practised widely in UK at the present time (It is sold under the trade name CERTAN).

5.11.1.2 Lesser wax moth (*Achroia grisella*).

This moth which is much smaller and has a wing span of about $^3/_4$ to 1 inch (19 to 25mm) and weigh about a sixth to a tenth of the weight of the greater wax moth. The female generally lays eggs in crevices and it is generally accepted that the larvae do not cause damage to the woodwork. However, we have heard reports to the contrary but we, ourselves, have never experienced woodwork damage with these lesser wax moths. The larvae still consume and digest the wax comb and while doing so they produce a large web of silk tunnels.

It is not so much of a pest in this country as the greater wax moth but can completely ruin comb if an infestation occurs and no protective measures are taken.

5.11.1.3 Death's head hawk moth (*Acherontia atropos*).

This is a magnificent looking moth with a skull and cross bones marking on its thorax on the dorsal side. It is attracted to bee hives and is quite rare in UK; it originates from N. Africa and Spain. It is worth having a look at one in a natural history museum.

It has been reported in Devon that more frequent sightings of these moths have been seen during 1998. Whether they are on the increase or whether they are taking advantage of weaker colonies during a poor honey season remains to be established.

The adults feed mainly on sap from tree wounds but they can resort to robbing nectar and honey from the hives of honeybees usually attacking the hive at night. They are not generally a pest in this country as can be noted by the *Acherontia* cadavers found occasionally in the hives of the honeybees.

5.11.2 Methods of storing comb with particular reference to prevention of wax moth damage and disinfection against nosema.

5.11.2.1 Types of comb to be stored: are supers and brood comb.

a) Supers. These can be stored either wet (with honey) or dry after being cleaned up by the bees and removed from the hive again. This comb consists only of wax and honey (if stored wet).

b) Brood combs. This type of comb is very different containing wax, pollen, larval skins and faeces, propolis, etc. making them much more attractive to attack by other insects and mammals.

5.11.2.2 The main causes of damaged comb are by;

a) Mammals such as mice, rats, squirrels, etc. which are easily excluded with travelling screens or queen excluders at the top and bottom of the stacks of boxes of frames.

b) Insects, the main cause of damage being the wax moths. There are two:

1. The lesser wax moth (*Achroia grisella*) and
2. The greater wax moth (*Galleria mellonella*) which is regarded as the major pest. However both can cause very extensive damage in a short time if precautions are not taken.

5.11.2.3 Methods of protection against wax moth damage.

There are four methods namely paradichlorobenzene (PDB), acetic acid (80%), heating and cooling. Each of these either kill (k) or have no effect (ne) on the various stages of development ie. egg, larva, chrysalis and adult. Reference to published literature reveals the following:

	PDB	ACETIC	FREEZING	HEATING
Eggs	ne	k	k	k
Larva	k	ne	k	k
Chrysalis	?	?	k	k
Adult	k	k	k	k

T (FREEZING) = 0°C to -17°C for a few hours to a few days depending on temperature and bulk of frames.

T (HEATING) = 120°F (49°C). Note the melting point of wax ≈ 145°F (63°C)

? = no reference could be found in the standard literature.

It will be clear from the above table that the best method is freezing before storage at normal temperatures. No airing of the combs is necessary. A good method for supers where the risk of disease is very low compared with brood frames.

Heating would also be possible but wax is very malleable at 120°F (49°C) and the temperature control would have to be precise.

Gamma radiation is known to kill all stages in the life cycle but is expensive and not really a practical method for the beekeeper. For the average beekeeper fumigation is the more usual method and it will be clear that both acetic acid and PDB is necessary and is the accepted method of dealing with brood frames.

•Brood frames. These should first be fumigated with 80% acetic acid as follows:

a) Brood box with frames placed on suitable board (acetic acid attacks concrete).
b) Metal runners in brood box well greased to protect from acid fumes. Any metal ends are removed from the frames.
c) An empty eke is placed over the brood box and covered by a suitable board.
d) The acetic acid is poured onto an old piece of rag in a shallow dish standing on the tops of the frames. The rag extends over the edge of the dish acting as a wick. The fumes are heavier than air and fall through the frames. The amount of acetic acid required is 100 ml per BS brood box.
e) All the joints should be sealed with tape to make the set up as air tight as possible.
f) Leave for one week and then the frames can be stacked for winter storage.
h) All the exposed wooden parts of the frames should be scraped clean of wax and debris before fumigation.

After fumigation with acetic acid they should be stacked as follows:

a) Mouse excluder with newspaper over.
b) Sprinkle one dessertspoonful of PDB crystals onto newspaper and place brood box over.
c) Cover with newspaper, PDB, another brood box, etc. finishing with a screen and crown board.
d) The PDB should not come in contact with the wax comb where it will contaminate the wax.

• Supers. It is unnecessary to fumigate with acetic acid and they can be stacked straight away with PDB and newspaper. At the end of the season it is important to get all frames cleaned up, fumigated and stacked for winter as soon as possible. While the weather is warm the wax moth can do considerable damage.

Note that if supers are stored wet it is necessary to make them bee-proof if they are stored outside otherwise they must be in a bee-proof shed or room. When the frames are wet, they are not attractive to wax moth so they can be stored without PDB. They do become very damp (honey hygroscopic) and tend to grow mould during the winter.

5.11.2.4 Other relevant points:

a) PDB crystals should never come into contact with the wax comb.
b) Disinfection with acetic acid is the approved method of cleaning comb infected with Nosema spores, so the storage treatment ensures that there is no risk of the spread of infection the next season when the comb is reused. Other pathogens are killed with acetic acid such as chalk brood fungus spores so it is good beekeeping practice to make this storage the norm.
c) Rather than have the trouble of stacking and sealing boxes for acetic acid treatment it is probably better to have a permanent installation to hold as many frames as required. It can be custom built and made completely air tight thereby using less acetic acid. The disinfection box automatically can become a frame storage box in the winter.
d) After any fumigation, combs should be well aired before re-use in the hive.

5.12 How mice and other pests can be excluded from the hives in winter.

The major pests requiring consideration for successful wintering are mice, other mammal pests, birds and human beings. Each will be examined in the following paragraphs.

5.12.1 Mice. These include the common or domestic mouse and the field and wood mouse. They will enter hives in the autumn seeking somewhere dry and warm to build a nest for hibernation purposes. This activity is prompted by the shorter days and a drop in temperature. The moral is to have mouse guards on the hive in plenty of time to ensure that the mouse does not enter.

• Mice feed on pollen, honey and bees. They therefore cause damage to comb, frames and hive equipment. In winter they will disturb the winter cluster and this disturbance can kill the colony if the temperatures are very low. The bees can sting mice to death and they have been known to be embalmed in propolis because the bees cannot eject them from them hive. Any droppings and urine are generally cleared out by the bees.

• Mice have oval skulls and can squeeze through a $^3/_8$ inch (9mm) wide slot but they cannot pass through a $^3/_8$ inch (9mm) diameter hole. Mice are therefore not a problem to keep out of the hive and if they do enter, it is the fault of the beekeeper not taking the necessary precautions in time.

5.12.2 Other mammal pests. These include shrews, rats, moles, squirrels, hedgehogs, etc. All these can disturb an overwintering colony and in this respect can cause damage to it, but many of them are hibernating themselves. We have noticed pronounced scratch marks at the entrance to some of our hives at one apiary and believe it to be due to badgers although we have not caught them in the act.

5.12.3 Birds as pests. The main culprit is the green woodpecker in very cold weather. They usually peck through at the hand-hold on National and Commercial hives and in a matter of an hour can make a hole of sufficient size to enter. If they are not spotted in time the colony will surely perish for it will occur in periods of hard frost or snow on the ground. Combs, frames and hive parts will be damaged.

Other birds are swifts, tits, swallows, shrikes, etc. taking bees on the wing (including queens on mating flights). We have watched sparrows in the early morning sitting on top of the hive waiting for bees to come out, catching them and taking them back to their nest for the fledglings. Pheasants also have a taste for bees; we wondered why one colony at one apiary was very often irritable until one morning we saw a pheasant tapping at the entrance and eating the bees as they came out to investigate.

5.12.4 Good ventilation while excluding mice. A colony during winter, if it metabolises 35lb (16kg) of honey, will be required to get rid of approximately 4 gallons of water. This can only be achieved by evaporation. The average rate is 5 pints/month or 3 ozs/day. It is more difficult for evaporation to take place in the damp western side of UK compared with the drier eastern side. These are the facts, the best configuration for achieving this evaporation is still being debated in the bee press and still no one seems to agree on the subject.

Our own method, which we have used successfully for many years after experimenting with various approaches, is as follows:

a) Heat escaping from the cluster causes the movement of air, warm moist air moves upwards and is replaced by cold dry air at the bottom.

b) All entrance blocks have 9 x $^3/_8$ inch (9mm) diameter holes drilled in them spaced equidistant apart across the length of the block. This gives a total cross sectional area of c. 1 square inch (650mm^2)to limit the air flow. The block turned through 90° is a normal reduced entrance block with a 4" wide slot. The $^3/_8$ inch (9mm) diameter holes form the mouse guard and are 'kinder' to the pollen collectors in the spring compared with the perforated metal strips sold by the equipment suppliers.

c) The crown board is raised about $^1/_8$ inch (3mm) with matchsticks at each corner. This gives an exit area for air to escape of c. 9 square inches(5800mm^2). The feed hole(s) are covered so that the flow of air is round the outside of the cluster and avoiding the chimney effect directly above the cluster. The roof ventilators now play no part in the ventilation system.

d) The mouse guards are put in usually in September before the ivy flow and the crown board is raised as late as possible to stop the gap being propolised. It is interesting to note that if there is no ventilation at the top of the colony, in the spring it is usually a mess with condensation and mouldy outside combs. One associated problem is that some of the stored pollen also develops mould and is then useless to the bees.

e) Strains of bee that produce a lot of propolis will get themselves into this situation if crown boards are raised too early.

5.12.5 It is very important to keep human beings out of the hive during winter or at the very least to stop them tinkering with it. No disturbance during the winter period is essential. Once the colony has settled down for winter it should be left undisturbed until the following spring. Experiments have been conducted and it has been found that the cluster temperature is raised quite a considerable amount (up to 10°F or 5.5°C) by say just taking the roof off. Such increases in temperature shorten the life of the winter bee and this manifests itself in spring just when the colony requires all the bees it can muster. Hives should never be sited under trees where the drip of water from the branches can cause colony disturbance.

5.12.6 Keeping the green woodpecker at bay.

The green woodpecker can spell disaster for a colony if they direct their attention to boring through the side of the hive. They are usually troublesome in very cold weather when they cannot find forage in the hard ground. There are two ways of protection:

   - Surrounding the hive with chicken netting ensuring that it is kept about 6 inch (150mm) from the hive walls to prevent the woodpecker from reaching through to the woodwork from the netting which provides him with a good foothold.
   - Covering the hive with a plastic bag (but note this interferes with the ventilation) and does produce condensation inside the bag. While it denies the woodpecker a foothold we do not recommend it because of ventilation and other problems.

5.12.7 Other points of interest are:

   a) It is desirable that a colony has stores of pollen which can then be used when brood rearing

128

starts after the winter solstice. We have never found this to be a problem but there are probably parts of UK where there is a dearth. The final topping up of pollen stores is during the ivy flow in September/October (nb. winter bees are produced by the consumption of large amounts of pollen). b) Plenty of bees are required for good wintering but making massive colonies by uniting can defeat the object as shown by some experiments done by Dr. E.P.Jeffree at Aberdeen University. The old adage that bees do not freeze to death but starve to death is very relevant to the wintering problem.

** ** ** **

APPENDICES

Appendix 1. Migration and evolution of the honeybee.

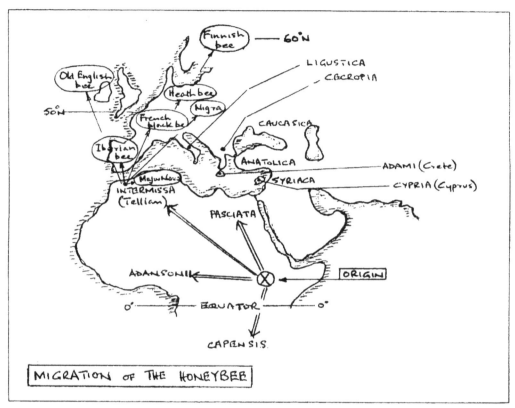

The above diagram shows the migration of the honeybee, from its origin in what is now Kenya, during the last 30 million years. All the major races shown by double arrows may be considered primary races and the one of importance to the UK situation is the *Apis mellifera intermissa*. It is from this bee that migrated northward into north west Europe that all the other sub-species have evolved, such as the French Black Bee and the Old English Bee shown with single arrows emanating from north west Africa on the diagram, ie. from the area of Morocco, Tunisia and Algeria. It migrated as far north as north latitude 60° close to the Arctic circle and became the Finnish Bee, a bee that is characterised by its ability to tolerate long periods of confinement in very cold conditions without cleansing flights. A truly remarkable evolution from an ancestor derived from the tropics.

The western honeybee, *Apis mellifera*, was transported to America initially by the pilgrims and to Australia and New Zealand by the settlers to these Antipodean countries. Until man took a hand in these migrations there were no honeybees in either of the two regions. The original bees were taken in skeps on the old sailing ships with a voyage time of about 6 months to the Antipodes; quite remarkable and a topic that would be an interesting line of research into the whys and wherefores of these exploits.

** ** ** **

130

Appendix 2. Food sharing in the honeybee colony.

Details of food sharing in a honeybee colony may be summarised as follows:

a) Food exchange (trophallaxis) is one of the most frequently observed behaviour patterns in a honeybee colony. It is going on 24 hours per day, day in and day out. Nectar or honey is passed from one bee to one or more receiving bees and is the prime mechanism for the exchange of pheromones within the colony (chemical communication).

b) Food is passed from worker to worker and from worker to the queen and drones, although the food provided for the queen is always royal jelly. Reciprocal feeding continues throughout the life of the bee. A bee up to 2 days old receives more than it transmits. The actual transfer starts by one bee either 'begging' or 'offering' food. Begging bees hold out their proboscis and offering bees fold back their proboscis and open their mandibles exposing a droplet of food.

c) During food transfer continual antennal contact takes place; the antennae of both bees are in continual motion touching one another. There is no known antennal language, the purpose is therefore somewhat obscure.

d) In addition to the exchange and distribution of pheromones with their chemical messages, the actual food transferred provides information concerning the availability of food and water within the colony. When no nectar is coming in, the bees use stored honey with a high sugar content which prompts water collection by the foraging bees. The bee can only metabolise sugars in a 50% solution; ripe honey contains 80% sugar. Conversely, when a heavy nectar flow is available (sugar content generally lower than 50%) the excess water requires to be evaporated and the water carriers are out of business.

Other information provided by food transfer is:
a) The type of nectar and pollen by taste and smell.
b) Water requirements for cooling; the house bees refuse to receive incoming nectar.
c) Wax building/secretion requirements; if there is nowhere to store the nectar it remains in the crop (honey sac) of the house bees.
d) Older bees (particularly foragers) are usually the ones offering food; this income stimulates the nurse bees to feed the queen and increases the egg laying rate of the queen.
e) Indicates to the whole colony the presence or absence of the queen.
f) The food transfer process is very rapid throughout the colony. A single transfer between two bees takes about half minute (transferring both food and pheromone). If these two bees feed two others, and then the four feed four others ad infinitum a series 2, 4, 8, 16, 32, 64, c.125, 250, 500, 1k, 2k, 4k, 8k, 16k, 32k, 64k is evolved; ie. 15-16 transfers each lasting about half a minute. In 7 or 8 minutes the whole colony is aware of the 'state of play'. This model is very simplistic and in practice the communication is faster than this. It is readily demonstrated by removing a queen from a colony. In about 5 minutes bees are busily searching the entrance area for her. In a matter of an hour or two emergency queen cells are likely to be started demonstrating the effect of shortage of queen substance.

** ** *** **

Appendix 3. Colony development during the active season.

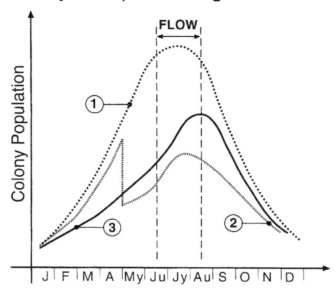

Graph 1. This illustrates a normal healthy colony building up on the spring flow to reach a maximum when the main flow occurs in June/July. The flow stops and the colony population then starts to decrease rapidly due to the queen's reduction in egg laying and the demise of those workers after their three week stint as foragers. The maximum population varies with the efficacy of the queen and the type of bee but in general may be assumed to be about 40,000 bees.

Graph 2. This illustrates a normal healthy colony which swarms in about April/May and then proceeds to requeen itself. It will be clear that it has a very much reduced colony population, and hence foraging force, when the main flow starts and therefore resulting in a very reduced honey crop. The amount of honey collected is directly proportional to the colony population.

Graph 3. This illustrates the build up of a diseased colony, say one with Nosema. It has a very reduced population and reaches its peak as the main flow is ending. Again this has a very serious effect on the amount of honey collected by the colony.

Considering the three graphs shown above it will be clear that in order to ensure the maximum honey crop it is necessary to have a disease free colony which builds up normally on the spring flow and does not swarm. It also illustrates the necessity of checking the colony for disease in the spring. If the colony is declared disease free and its performance is akin to Graph 3, then it is more than likely that the queen is failing and needs replacement.

** ** ** **

132

Appendix 4. Honey usage diagram.

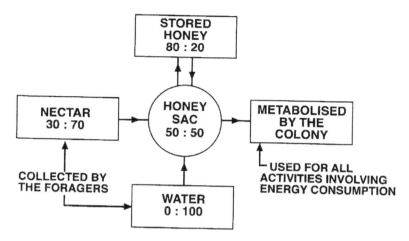

SUGAR : WATER RATIOS

Points of interest on the diagram:

a) The honeybee always tries to maintain a sugar/water balance of 50/50 in the honey sac.

b) When nectar is flowing into the hive, no water is required; manipulation ('ripening') loses approximately 15% of the water. Excess nectar, above colony requirements, is stored as honey after full ripening (evaporation).

c) When no nectar is flowing into the hive then honey reserves must be used and water is required for dilution of the honey from 80/20 to 50/50. Foragers become water carriers.

d) Similarly, if the hive is becoming too hot and fanning is not lowering the temperature sufficiently then house bees will refuse to accept nectar and foragers become water carriers.

** ** ** **

133

Appendix 5. The BBKA examination system.

The diagram below shows the system of examinations set up by the BBKA Examinations Board. It is necessary to pass the Basic Examination in order to proceed to all the other examinations.

BBKA EXAMINATIONS

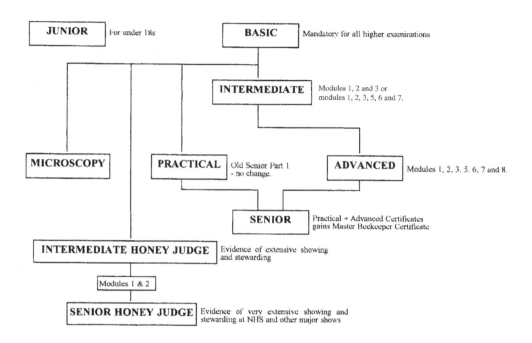

NB. BBKA Certificate in Husbandry. Candidates who dislike written exams can, after passing the Basic, take an oral Certificate of Husbandry after 3 years of beekeeping. Contact the Board for details (given in our Purple Book).

Modules
1) Honey bee management
2) Honey bee products and forage
3) Honey bee diseases, pests and poisoning
4) Intermediate biology
5) Honey bee biology
6) Honey bee behaviour
7) Selection and breeding of honey bees
8) Honey bee management and history

** ** ** **

Appendix 6. Additional items to cover the Scottish Basic Beemaster Certificate.

The Scottish Basic Examination is identical to the British Basic Examination except for 3 additional items which are as follows using the Scottish syllabus numbering:

3.14 The candidate should be able to describe the preparation of colonies for a particular honey flow.
3.15 The candidate should be able to describe methods of securing stocks prior to moving.
3.16 The candidate should be able to state the risks in transporting live honeybee colonies.

We have addressed these three topics in the following paragraphs.

A6.1 The preparation of colonies for a particular honey flow.

We must admit that we had some difficulty with this section of the Scottish syllabus mainly because of the vagueness of the wording. Special colony management is required for very early honey flows and those at the tail end of the season, for example, rape blooming in April and heather blooming in the last few weeks of August. All honey flows require the maximum bee population in each colony in order to take advantage of the available nectar; maximising the colony population for an April flow requires very early seasonal management not normally associated with the normal colony spring build up. Our other problem was whether migratory beekeeping is inferred or whether it is a particular honey flow associated with a fixed apiary site. We have assumed that fixed apiary sites are involved rather than migratory beekeeping which is more often the preserve of the commercial beekeeper.

A typical season for both honey and pollination would start with plums followed by pears, cherries, apples, rape, sycamore, horse chestnut, strawberries, raspberries, beans, lupins, mustard, white clover, bramble and finally heather to name some of the major sources. The flowering period of some of the above will overlap and not all hives will be working the same crop at a particular time.

A6.1.1 The basic requirements.

To manage colonies for maximum honey production there are some basic requirements which need to be observed irrespective of district, weather conditions or time of flowering of the forage plants. These requirements are discussed in the following paragraphs.

The starting point must be the seasonal colony population cycle (see Appendix 3). It is applicable in principle to our temperate zone in the United Kingdom and is valid for latitudes from about 35°N to about 65°N. It is based on a colony build up in the spring flow peaking in the summer (summer solstice) to take advantage of the main flow. It assumes that the colony does not swarm. The length of time that flows occur are quite long compared with higher latitudes. The graph of the colony population cycle is applicable to our native honeybee, another basic requirement where early flows are concerned. The yellow Italian bees work best in warm temperatures.

Listing the basic requirements we have:

- The type of bee.
- Good manipulative skills to prevent any swarming.
- Maintenance of disease free colonies.
- Queen selection for breeding purposes.
- Maintain all colonies virtually identical to enable them to be managed as identical units.
- All units must be as large as possible in bee population.

Very few beekeepers can meet the above requirements. Those who do may be compared to those gardeners with 'green fingers' whereby their plants always seem to succeed; in the beekeeping world the likes of Bro.Adam immediately come to mind. Such ability results in the colonies not swarming, colonies with maximum populations throughout the year which are then prepared to take advantage of any flow when it starts. Ideally the colony should be used in the most economic and cost effective way after the flow.

A6.1.2 The bees.

Early flows means that the bees being used for these flows must be available early in the year in colonies as large as possible and they should be able to work at low temperatures. Bees from the tropics and sub-tropics, ie. the Mediterranean area, do not work in low temperatures and tend to have a slow colony build up in the chilly spring weather, waiting until it gets warmer before brood rearing starts with a vengeance.

It will be clear that a bee more suited to our climate is a prerequisite for this work and, of course, this means the black bee of European origin. Thus it is essential to be able to breed a bee suitable for the purpose and this has to be done from the stocks locally available to the beekeeper selecting for best honey yields and temper. The only way this can be done is by keeping good records and analysing them at the end of each year. Then it will be clear immediately which stocks are giving the hundredweights of honey and which are producing swarm cells in the spring.

Importing queens from other sources in the UK is no guarantee that the queens will be satisfactory and time will preclude experimentation in this direction as well as the expense involved. Unfortunately it has been our experience that the majority of beekeepers do not find the time for good queen rearing practice and while their bees produce average crops of honey, the bees are generally very bad tempered.

On the other hand, if early crops are not a pre-requisite then other races of honeybees can be entertained again with the objective of good temper and low swarming propensity.

A6.1.3 The prevention of swarming.

If advantage is to be taken of a particular crop(s) during the active season then it is imperative that the colony does not swarm. If it does then about 50% of its population will be lost with a similar reduction in the foraging force. Reference to the graph in Appendix 3 makes this abundantly clear when compared with the colony that doesn't swarm (ie. the annual population cycle for a non-swarming colony). With the best will in the world a colony will make preparations for swarming, no matter how well it is managed and it is then incumbent on the beekeeper to prevent the loss of bees by adequate methods of swarm control.

A6.1.4 Maintenance of disease free colonies.

Reference to Appendix 3, the graph shows the build up of a diseased colony which may be compared with a normal healthy colony (annual population cycle). It will be obvious that a diseased colony in the spring is likely to be a useless honey producer unless action is taken early in the year. This demonstrates the wisdom of monitoring all colonies for disease regularly; autumn and spring sampling for the adult bee diseases and at appropriate times for the brood diseases.

Whenever colonies of bees are moved they are put under stress, increasing the risk of weak colonies developing Nosema, EFB, etc., the well-known stress diseases associated with many commercial beekeeping enterprises. We do not think that there is a solution to this problem if the bees are continually on the move. Continual colony assessment can be the only answer coupled with hive hygiene but here it must be appreciated that this will increase the labour costs of the operation which are already likely to be running on a tight budget.

A6.1.5 Selective queen rearing.

The provision of young queens in all stocks is a necessary preparation for a productive colony to make best use of the flow. Selective queen rearing is outside the scope of this syllabus but nevertheless candidates should take notice of this requirement; it is addressed in the modular examinations which follow this level of study.

A6.1.6 Maintaining large stocks.

Maintaining large stocks clearly maximises the work force available to take advantage of any honey flow when it occurs during the season. In essence this means having a young prolific queen in a disease free colony which doesn't swarm, all of which have been discussed above.

A6.1.7 Preparing the stock for an early flow in April, eg. rape.

The objective in preparing the stock for an early flow is to endeavour to have the largest colony possible by the time the rape comes into bloom and starts yielding nectar. Brood rearing has to be stimulated and to do this pollen is required. Brood rearing only starts with a vengeance when the foragers can collect fresh pollen. As brood rearing requires to be stimulated from mid January, when virtually no pollen is available, it is necessary to feed what are called, ' pollen patties'. One word of warning before preparation starts; do check with the grower which variety of rape is being grown as there are now types which produce virtually no nectar. We have had apiaries in recent years with rape the other side of the hedge and the bees have not collected a pound of honey.

A6.1.7.1 Preparation of patties.

If patties are to be successful they must support brood rearing, which may be stating the obvious but many substances have been tried and tested before arriving at the most suitable. For example, in Canada fish meal has been tried and is cheap and successful. All are mixed with sugar syrup which not only enhances the making of the patties but also provides sugar which is acting to some extent as a phagostimulant. As a result the standard recipes for patties in the UK has evolved to;

Ingredient	Pollen substitute patties	Pollen supplement patties
Fat-free soya flour (%)	75	60
Brewer's yeast (%)	25	20
Natural pollen (%)	0	20
Totals by weight (%)	100	100

The whole is mixed into a stiff paste with sugar syrup and made into patties of weight between ½ and 1 pound with a thickness of about ½ inch when spread on to plastic film or waxed paper and stored in plastic bags to stop them drying out.

Pollen substitutes are made commercially and sold under enticing names; they are no better than using the above which will prove to be much less expensive. Since the late 1960s we are not aware of any further work which has been undertaken on artificial materials to replace pollen. Generally, our knowledge is extremely meagre in this direction.

Bees, if given the choice, will select pollen supplement cf. pollen substitute. Both can be made more attractive to the bees by adding small amounts of fragrant scents including honey essence; so perhaps the moral here is to mix the patties with run honey rather than sugar syrup, providing the run honey is from a known source and is not of foreign origin with the likelihood of it containing AFB spores.

We must emphasise the importance of using fat-free soya flour rather than ordinary soya flour which has a high fat content and is unsuitable for bees. We know of one case where a nucleus succumbed after being fed pollen substitute patties and the only reason we could determine was the doubtful source of the soya flour. Note that expeller process soya bean flour = fat-free soya bean flour.

A6.1.7.2 Use of supplements and substitutes.

The patties are messy to make but very easy and simple to use. The patties are put directly onto the brood chamber frames with the plastic film uppermost. It pays to have the plastic much bigger in diameter than the patty to prevent it drying out and becoming crumbly; the bees seem to turn their noses up at the dry bits.
In the UK, patties are usually used in the spring in areas which are predominantly rape growing where there is a dearth of spring flowers to allow the colony to build up naturally. After lifting the crown board, a puff or two of smoke will send the bees down and it easy to place the patties on top and press them down in between the seams of the combs. In February/March a healthy colony may be expected to consume a one pound patty in one week.

In order to build up the colony for the early spring rape flow the patties can be put on the colonies as early as the middle of January; the earlier the better to get the colony as strong as possible before the rape comes into bloom. The major drawback is that swarming problems are likely to occur earlier with such stimulation.

We have no experience of using pollen patties at other times of the year, eg. a flow associated with honeydew in July. There would be difficulties feeding the patties with supers on and we would not recommend feeding them at the entrance at that time of the year because of robbing problems.

A6.1.8 Preparing the stock for a late flow in August, eg. heather.

Stocks destined for the heather in August have been used for the normal season up to and including the main flow. At this time of the year the strength of the colony is naturally declining quite rapidly, see Graph 1 in Appendix 3. It is normal practise to requeen the stock with a current year queen to encourage egg laying to continue for as long as possible to ensure maximum foraging force before moving the stock to the heather.

The brood chamber should be as full of brood as possible with food on the two end frames of the brood chamber to ensure that the colony does not starve if it is prevented from foraging during its first 10 days on the moor. Ideally the flow should be just starting when the stock arrives on the heather and, because the brood chamber is chock-a-block with stores and brood, honey is stored immediately in the super provided.

If the flow is good then the stock is likely to provide a surplus after providing enough winter stores for itself in the brood chamber.

A6.2 Methods of securing stocks prior to moving.

A6.2.1 Preparing a stock for moving: starts by removing the crown board and replacing it with a travelling screen, preferably with a space of about 1 inch on the underside to allow room for any bees to cluster. When being moved, the entrance should be closed (eg. reduced entrance block with foam pushed into the reduced entrance just before moving) and not restricted with a screen as many books recommend. If light is showing at their normal entrance they will attempt to escape at this point and there is the danger of them suffocating in the panic to get out. When being moved, the hive parts have to be secured one with another; this can be done in a variety of ways:

a) Using a hive strap around everything excluding the roof, which is always removed for travelling. Two hive straps in opposite directions are safer than one.
b) Screwing plates (4in x 1in) at an angle of 45° across the joins between floor and boxes and the screen being fastened with screws to the top box. Note that the 2 plates on each side should be angled in opposite directions (so the plates are at 90° to one another) to prevent movement. This method is considered to be superior to all others but it is more time consuming. It should be used for a major move over long distances, say greater than 50 miles.
c) Spring clips to join the boxes together: these use 3 screws, 2 on one box and 1 on the other.
d) Bro. Adam's method of long bolts through the screen, brood chamber and floorboard.
e) Using hive staples; these are a bit outdated these days and a fine way of disturbing a colony

when hammering them home.

f) The entrance block needs securing to the floorboard, the safest way is with two 'L' brackets screwed to the front and to the sides of the floorboard.

A6.2.2 Other preparations which are necessary before the actual move are as follows:

a) The site and stands at the far end should be ready to receive the stocks immediately on arrival.

b) Prepare emergency equipment for journey, ie. veil, smoker, fuel, water spray for occasional cooling, spare ropes, wide sticky tape for accidental bee leaks, etc.

A6.3 The risks in transporting live honeybee colonies.

A6.3.1 Moving the stocks: involves observing some simple rules:

a) Place the foam in reduced entrance and then remove roof.

b) Place the stocks in the transporting vehicle with the frames in a fore and aft direction so that frames cannot swing if emergency braking or stopping is required en route.

c) Ensure that all stocks are roped down securely before starting. Stop after 15 minutes and check all is secure (tension up if required).

d) Corner at slow speed to minimise frames swinging.

e) The stocks should be moved preferably during the hours of darkness arriving at the destination about daybreak. If they are moved during the day over-heating must be watched carefully and cooling applied (say every hour) with water spray, if necessary, through the top screen.

f) If they are being moved on a trailer ensure that it has a spare wheel.

g) On arrival set up all stocks in final positions, replace all roofs and then remove foam at reduced entrances quickly.

h) Next day remove travelling fixtures and screens and replace crown board nb. leave joining plates in situ for the return journey. Some beekeepers who are regularly moving stocks of bees have their floor boards permanently screwed to the brood box.

i) Depending on the weather and the strength of the stock/s it may be prudent to add a queen excluder and a box of super frames once the travelling screen has been removed. See section 4.6 on supering.

A6.3.2 The criteria to be observed when moving colonies of bees from one place to another (including optimum distance, vibration, temperature, ventilation and water supply).

A6.3.2.1 The distance: that bees can be moved is well known, ie. 3 feet maximum or 3 miles minimum, if no bees are to be lost from the colony concerned. The reason for the distance restriction is twofold. Honeybees forage generally up to a distance of $2^1/_2$ to 3 miles from their hive and have a 'mental picture' of this area or recognise distinctive landmarks within the area and know how to navigate back using these landmarks. Moving their hive within this known area creates a condition whereby the foragers leave the hive in the new position, re-orientate on leaving the hive but while foraging, recognise well known landmarks and return to the old site. The navigational ability of the honeybee is extremely precise (a matter of a few inches near their own hive). Moving the hive entrance more than 3 feet will create a condition whereby the foragers will not find their hive and will either drift to a nearby hive or cluster at the original position of the hive entrance. The authors conducted a series of experiments some years ago to test the memories of the bees by moving them to a distant apiary and then returning them to a different site in the original apiary. After about two weeks their memories seemed to fail and all foragers returned to the new hive position in the original apiary. For periods away from the base apiary for less than 2 weeks then the stocks of bees should be returned to their original site when brought back. Of course during the 2 week period many of the original foragers would have died a natural death and new foragers would have taken their place.

A6.3.2.2 Vibration: excites bees and if they are closed up in transport the temperature increase would be dangerous if insufficient ventilation and cooling were not provided. During transportation by vehicle there will be a continuous vibration keeping the colony in a state of agitation and high temperature. It will therefore be clear that vibration in general is closely allied to temperature and ventilation. In order to minimise these adverse effects, stocks should be handled with care during the loading and off loading process.

A6.3.2.3 Temperature and ventilation: go hand in hand and, of course, are allied to vibration. Because of the rise in temperature when a colony is disturbed it is necessary to provide adequate ventilation when moving bees. If very strong colonies are to be moved then it may be advantageous to provide additional space by adding another super as well as providing the ventilation screen. Even with these precautions moving strong stocks during the day in warm weather may be insufficient to prevent dangerous temperature rises, enough to melt wax comb and drown the bees in honey. Spraying the colony with water through the ventilation screen will be required as part of the operation.

A6.3.2.4 Water supply: Water may be required en route as indicated above and it can also be provided by placing a sponge fully absorbed with water in an empty super over the colony and below the travelling screen.

A6.3.2.5 Other points: related to moving bees are:

a) Bees should only normally be moved during the flying season, the winter cluster should not be disturbed.

b) Continual movement of bees, for say pollination purposes, puts them under stress and stress is the forerunner to nosema and EFB.

c) It is better to move a stock of bees some days after it has been inspected in order to allow time for the bees to re-propolise all the seals which had been broken. This minimises internal movement of frames etc.

d) Travelling screens should be constructed of a mesh of 7 to 1 inch in a wire gauge of c. 28 SWG.

e) It will be obvious that a regular water supply will be required by the colony when it arrives at its new location. A point which is often overlooked.

f) Stocks of bees are best lifted by two people. Not many beekeepers have mechanical hoists but many suffer from backache. Trolleys and wheel barrows designed for moving hives are available on the market but a DIY sedan type hive transporter (stretcher model) using a beekeeper fore and aft is best when the terrain is rough.

** ** ** **

141

Appendix 7. Additional items to cover the Federation of Irish Beekeepers' Associations Preliminary Certificate.

The Preliminary Examination of the FIBA is identical to the British Basic Examination except for 3 additional items which are as follows:

1. The use of nectar and honey in the life of the colony.
2. The use of the queen excluder.
3. The hiring of colonies for pollination services.

These three syllabus items are addressed in the following paragraphs.

A7.1 The use of nectar and honey in the life of the colony.

Nectar and honey provide the carbohydrate part of the honeybee's diet thereby providing for all its energy needs which are considerable. Foraging (flying) is the major item in a long list. Naming a few others, we have walking, fighting, wax making, comb building, heating the nest, ventilating by fanning, etc.

The use of nectar and honey during the course of a year is enormous and is used mainly for foraging, brood rearing and wax making; it is never seen by the beekeeper. Any surplus collected will be in excess of the amount shown in the calculation below for these three items. It will be clear that the foraging activities represent a large percentage of the overall expenditure and the larger the colony the higher the figure is likely to be. Note the difference in the prolific Italian yellow bee that continues rearing brood irrespective of weather conditions and produces enormous colonies compared with the more thrifty black bee of northern latitudes. The balance is redressed when the flow starts and there is a much larger foraging force to utilise it to advantage. The large colony of yellow bees usually wins hands down on surplus honey and repays the beekeeper's sugar bill for syrup required in times of dearth.

• The annual calculation is shown below for some of the honey that the beekeeper never sees.

Honey used by a flying bee = 10mg/hour
Amount used by a forager flying for 5 hours/day = 50mg
If the queen lays an average of 1000 eggs/day for 100 days = 100,000 eggs
The honey required to rear one bee = 50mg (we have seen figures as high as 100mg)

• The annual honey consumption can be calculated as follows:

Foraging: (100,000 bees × 50mg/day ×21 days) ÷ (1000 × 454)   = 231 pounds
Brood rearing: (100,000 eggs × 50mg) ÷ (1000 × 454)   = 11 pounds
Wax making: 1 pound of new wax per active season   =   8 pounds
TOTAL before considering surplus   = 250 pounds

142

In general the foraging behaviour may be summarised as follows:

a) The number of foraging trips that a worker makes per day is between 7 and 13 with the maximum about 24 and an agreed average of 10 per day.
b) The average load is between 40 and 50mg.
c) Time foraging per trip = 27 to 45 minutes.
d) Time unloading and dancing in the hive = c. 4 minutes.
e) Number of visits to different florets = c.100 and depends very much on the flow and the flower being worked.

A7.2 The use of the queen excluder.

The purpose of a queen excluder is to exclude the queen from the supers while allowing worker bees access to them thereby keeping all the brood rearing and associated pollen in the lower area of the hive, ie. the brood chamber. Theoretically only honey would be stored in the supers but in practice it is found that the first super often has quite a few cells with pollen stored in them. There doesn't seem to be any answer to this if the brood nest extends close to the top of the brood frames because the bees will always store pollen directly adjacent to the brood where it is required for use. The queen excluder is attributed to Abbé Collins in France in the year 1865.

The general principle is a flat sheet with slotted holes just large enough to allow a worker to pass through (it not only prevents the queen passing through but drones as well). Zinc sheet is a popular material; the size of the slots was 5/32 inch (4mm) or 0.157 inch but most slotted types are now made with slots of 0.162 inch or 0.163 inch depending on which book is read. The same effect can be obtained with a grill of parallel wires.

A7.2.1 Slotted types: Generally made of zinc but galvanised mild steel slotted excluders are now available. They were made in two versions, one with a series of short slots (c. $1^1/_2$ inch) and the other with long slots (c. 3 inch). The design is for bottom bee space hives to allow the flat sheet to lay directly on top of the frames. The long slot variety was easily damaged and the short slot version is generally preferred. It is possible to frame this type of excluder as a DIY job; they are not available commercially in a frame. They are the cheapest of all excluders to buy. The mild steel short slot version is probably the best of those available. In recent years a slotted type has been available made from plastic; to date we have no experience with this type of excluder.

A7.2.2 Wire types: have all to be constructed with strong rigid wires to prevent damage and bending of the wire during use. The construction must be able to withstand damage by burr and brace comb when it is removed from the hive during manipulations. The gaps between the wires must not be greater than 0.165 inch. All wire type excluders are framed and should have a bee space on one side only; some are on the market with a bee space on both sides which is wrong and they should be avoided. The framed wire type is known as the Waldron excluder and similar types from Germany are known as the Herzog excluder. Both are more expensive than the slotted types. A further type with wood/wire/wood construction is available in USA at an even greater cost; it is claimed the bees 'like it better'(?) than other types. Better ventilation through the hive is achieved with the wire types compared with the slotted types.

A7.2.3 Other points of interest:

- If the excluders are electroplated with zinc (and most are) they should be cleaned with boiling water or the careful use of a small blowlamp. They should not be scraped, the plating will be damaged and rusting will occur.
- When replacing an excluder, ensure that the top bars are clear of brace comb which may distort the excluder or, worse, damage it.
- It is quite amazing how often the excluder is put on the wrong way round, the bee space should be below on a bottom bee space hive. Which way should it be placed on the frames; parallel to the frames or at right angles to them? In practice this does not matter as there is a bee space on both sides of the excluder when in use.
- At one branch meeting the authors saw a Waldron type excluder jammed full of dead worker bees stuck in the slots between the wires. It turned out that the beekeeper had been sold an excluder for *Apis cerana* which is a smaller bee cf. the *Apis mellifera* - most unusual!

A7.3 The hiring of colonies for pollination services.

A7.3.1 General. It is well known that most crops require to be pollinated by some means or another; generally, the honeybee can provide this service and it can provide it most efficiently if the colonies are managed specifically for pollination services. This means improving the location of the honeybees' nest site with respect to the crop. In the case of honeybees this means moving the stocks into or adjacent to the crop to be pollinated, for example onion seed plants, orchards, etc. All crops are improved both in quantity and quality by efficient pollination and many beekeepers provide a pollination service to growers by hiring their colonies to the grower for a short period while the crop is in blossom.

A7.3.2 An appraisal of the variables involved.

Successful pollination and fertilisation is dependent on a number of factors:

a) The availability of an adequate number of pollinators (eg. bees).
b) The weather conditions to allow the pollinators to fly.
c) The temperature to produce nectar to attract the pollinators.
d) Humidity to be relatively low so that the pollen is viable; often large
e) The temperature must be high enough to complete the pollen tube growth before the flower aborts. quantities of dead pollen are transferred by the bees to no avail.
f) Some species require cross pollination.

The beekeeper and the grower will have no control over the weather and both have to accept what comes. If the temperatures are high enough to activate the enzymes controlling the secretion of nectar from the nectaries of the flowers on the crop they may also be high enough to activate those of a more attractive source of forage for the bees, such as dandelions beneath the trees in the orchards. This problem must remain that of the grower, who can have them cut down. One quoted measurement we found indicated that the nectar sugar content of dandelion was 73% while that of the pears which required pollinating was 8 to 10%. Often the nectar sugar content of top fruit can be low and timing

144

the importation of colonies of honeybees on to the crop is critical to ensure that they work only the crop.

A further problem occurs, particularly with apples, where cross pollination is required; it is important to have the layout of the trees in a pattern that ensures the bees will forage on both main crop and pollinators. This is a grower responsibility and beyond the control of the beekeeper.

The question that next arises is how many colonies and how should be sited throughout the orchards. If the bees in a colony can fly as far as 2.5 miles in search of forage, then they have an area of c.12,500 acres at their disposal. The bees are likely to be spread rather thinly unless other colonies are in the vicinity. What then are the requirements for say a 20 acre orchard requiring pollination services? There is no mathematical way of calculating these requirements, as far as we know. By experience over the years a density of 1 colony per acre (or 2 per hectare) has been found to be satisfactory for apple and other top fruit. Therefore our 20 acre orchard would need about 20 colonies to fulfill a pollination contract.

The final question is how should the colonies be distributed throughout the orchards? We have not been able to find much meaningful information on this aspect but the consensus of opinion seems to be that the colonies should be spread throughout the crop.

A7.3.3 Arrangements with the farmer grower.

Whenever a beekeeper provides stocks of bees for pollination purposes to fruit growers, seed growers, etc. it is important that there is agreement between them in the form of a written contract to prevent misunderstandings and future legal wrangles. The grower has certain requirements and the beekeeper must be quite clear about his obligations. Below is the basis for a typical contract which may be varied to suit individual requirements. It is divided into two separate parts, namely, the grower and the beekeeper.

Agreement made on/date/between
The Grower (by name) of the one part of/address and telephone number and/
The Beekeeper (by name) of the second part of/address and telephone number
agree to enter into a pollination contract in respect of/ (name of crop) located at address of/
orchard/ farm, etc.

The Grower agrees:

a) To give.........days notice to bring the stocks to the crop.
b) To give.........days notice to remove the stocks from the crop.
c) To pay a fixed fee of..........when the stocks of bees are delivered.
d) To pay a rental of ....per stock per day while the stocks are on site at the crop.
e) To pay a fixed fee of ....when the stocks are removed plus the rental due.
e) To pay 1 per centum plus the current bank rate per calendar month or pro rata on amounts unpaid after the due date.
f) To use no pesticides, fungicides or insecticides on the crop during the agreed rental period

except with the agreement of the beekeeper.

g) To provide an uncontaminated and continuous water supply during the rental period for the bees; details to be incorporated.

h) To assume liability for the stocks in respect of vandalism.

i) To indemnify the Beekeeper against 3rd party claims in respect of stinging for the duration of the contract.

The Beekeeper agrees:

a) To place the stocks in the pattern agreed selected by the Grower. Include a sketch of the layout as a schedule to the agreement.

b) To leave the stocks on the crop for an estimated period of..............days, any extension of this time shall be by mutual agreement between the two parties.

c) To open the stocks and demonstrate the strength of the colonies and the amount of brood present as required by the Grower; any one colony only being opened for inspection once. Each colony shall have for charging purposes..............frames containing brood, which may be sealed or open, containing eggs and larvae.

d) To manage the stocks during the rental period and maintain them in good condition in accordance with good beekeeping practice.

Signed...Grower..(date)

in the presence of...Witness

address..

and

Signed...Beekeeper..(date)

in the presence of...Witness

address..

** ** ** **

Appendix 8. The SBKA examination structure.

The structure of the Scottish Beekeepers' Association examination system is shown in the diagram below. It is similar in many respects with that of the BBKA.

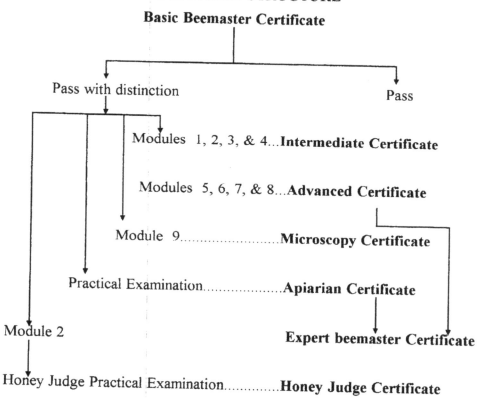

THE SCOTTISH BEEKEEPERS' ASSOCIATION

EXAMINATION STRUCTURE

Basic Beemaster Certificate

Pass with distinction   Pass

Modules 1, 2, 3, & 4...Intermediate Certificate

Modules 5, 6, 7, & 8...Advanced Certificate

Module 9.........................Microscopy Certificate

Practical Examination...................Apiarian Certificate

Module 2   Expert beemaster Certificate

Honey Judge Practical Examination.............Honey Judge Certificate

** ** ** **

Appendix 9. Additional reading extracted from the BBKA book list.

It has been suggested that we provide a guide for further reading; we can do no better than reproduce the first part of the BBKA book list dealing with the Basic examination and general reading for all the other examinations. This is as follows:

BASIC EXAMINATION

Guide to Bees and Honey	Ted Hooper	Blandford
Beekeeping Study Notes (Basic examination)	J.D. & B.D. Yates	BBNO
Beeway Code	D.A.R.G.	D.A.R.G.

GENERAL READING FOR ALL EXAMINATIONS

Beekeeping Study Notes (For all other modules)	J.D. & B.D. Yates	BBNO	
The Honey Bee	Gould/Gould	Scientific American	
Guide to Bees and Honey	Ted Hooper	Blandford	
Beekeeping - a Seasonal Guide	R. Brown	Batsford	
Beeway Code	D.A.R.G.	D.A.R.G.	
Handbook of Beekeeping	Dr. H. Riches	N.B.B.	
The Hive and the Honeybee	Dadant	Dadant	
World of the Honey Bee	C.G.Butler	Collins	o/p
Bees of the World	O'Toole/Raw	Blandford	o/p

o/p = out of print.

Each winter when we run our Basic course of tuition at the Swarthmore Adult Education Centre, Mutley Plain, Plymouth, in addition to the three recommended books shown above we include "The World of the Honey Bee" by C.G.Butler. Not only is it very instructive for those new to the honeybee, it is a good read. We have not yet found anyone who disputes these findings.

** ** ** **

148

INDEX

W

waistcoat pockets, 34
Waldron excluder, 143-4
warming cabinet, 96-7
water, 7-8, 18-9, 33, 36, 39, 41, 43, 45, 59-63, 67-8, 70-2, 83, 91-3, 95-7, 103, 105, 127, 131, 133, 140-1, 144-5
water carriers, 33
water collecting, 42, 44, 64, 67, 131
water supply, 51, 64
wax, 6, 9, 12-3, 33, 35, 45-6, 89-90, 105, 125, 141
wax building, 47, 67, 73
wax cappings, 8, 17-20, 31, 41-3, 92-4, 103, 106
wax comb, 7, 15, 35, 50, 141
wax extractor, solar, 104-5
wax extractor, steam, 104
wax foundation, 26-8, 105
wax glands, 16, 33
wax making/comb building, 34, 142
wax moth damage protection, 117, 123, 125-6
wax moth, bacteriological control, 124

wax moth, greater, 124-5
wax moth, lesser, 124-5
wax pockets, 34
wax production, 34, 67
wax scale, 34-5
wax secretion, 35
WBC, 24
Weed, 28
Weights and Measures Acts 1963 to 1979, 100
wet supers, 66, 71, 89
wings, 6, 15-7, 20, 49-50, 112-3, 116
winter bee, 9, 31-2, 43, 49, 62-3, 128-9
winter cluster, 127, 141
winter, preparations for, 18-9, 22, 43, 49, 63-4, 67-8, 70
wintering of a honeybee colony, 7, 9, 44, 48-50, 59
wintering, requirements for successful, 9, 43, 49, 53, 59, 60-2, 70, 114, 127-9, 139, 141
wiring, 15
wiring foundation, 15
work undertaken by worker bees, 33

Y

year's work in the apiary, 59
yeast, 138
Yorkshire spacers, 29

Z

Zander, 114

** ** ** **

153

Lightning Source UK Ltd.
Milton Keynes UK
UKHW03f0131160518
322662UK00005B/169/P